BIOGRAPHY OF A GERM

BIOGRAPHY

OF A

GERM

Arno Karlen

PANTHEON BOOKS   NEW YORK

Library of Congress Cataloging-in-Publication Data

Karlen, Arno.
Biography of a Germ / Arno Karlen.
p. cm.
ISBN 0-375-40199-7
1. Borrelia Burgdorferi.   2. Lyme Disease.
I. Title.

QR201.L88K37   2000
579.3'2—DC21   99-057304

www.pantheonbooks.com

Book design by M. Kristen Bearse

Printed in the United States of America

First Edition

2   4   6   8   9   7   5   3   1

Contents

CONTENTS

BIOGRAPHY OF A GERM

ONE

A Very Small Life

TO THE NAKED EYE, it is invisible, a nothing. Under the microscope, it seems a silvery corkscrew undulating on a dark field. The form has simple elegance, like the whorl of a nautilus shell or the sweep of a dragonfly wing. But that simplicity is an illusion. Through the more powerful electron microscope you see not a featureless wiggle but a shape-shifter—now a spiral, now a thread, now a rod or a sphere—with two walls, a dozen whiplike appendages and internal structures. And beyond any microscope's view, revealed only indirectly, by laboratory tests, lies a marvel of complexities. The surface bristles with molecules that sense and respond to the environment, and the interior churns like a chemical factory. Inside, more than a thousand genes flicker on and off in changing sequences, to allow survival in places as different as a tick's gut, a dog's knee and a human brain.

It is the bacterium *Borrelia burgdorferi*, by human standards a very small, brief flicker of life. Yet the boldest writer of science fiction could not invent a creature so ingenious, whose existence is entwined with that of so many other species. Although this microbe inhabits much of the earth and myriad hosts, it was not discovered until 1982, and then only because it had ignited a new epidemic, Lyme disease. That illness, so troubling to humans, is just a short, recent chapter in the germ's long history, and from its own perspective not the most important one. *Borrelia burgdorferi* has an ancient lineage, far older than ours, and despite all the vaccines and antibiotics we devise, it has a more promising future. It preceded people on earth and will doubtless survive us. For that reason alone it deserves respectful biographers.

Clearly there is much drama in this little theater. But that should be no surprise, for just as every person's life, seen close up, is compelling, so is every other creature's. *Borrelia burgdorferi* is proof that if you want to see life afresh and be struck with awe, you need only take a germ's-eye view of the world.

TWO

A Subject Not Picked at Random

LIKE MOST BIOGRAPHIES, this one will describe a life and its setting in time and place. It will tell about the subject's ancestors and relatives, its ties with friends, its struggles with foes, and its mark on other lives. This seems to me the usual task of biography, but a few readers will probably demur. However much I claim that *Borrelia burgdorferi*, though faceless, is unique and fascinating, they will ask, "But why a full-length, full-dress life of a germ?" The question presumes that biography has proper subjects and less proper ones. That is a narrow view of germs and of biography.

Every biographer has a purpose; whether modest or grand, conscious or taken for granted, it guides his choice of subject and his presentation. That is quite clear when you look at just a handful of the best practitioners. Plutarch, the first great biographer we know by name, meant his *Lives* of noble Greeks and Romans to show how character reveals itself in public life. The result was a didactic portrait gallery, illustrating moral choice in action. He achieved his aim so splendidly that two millennia later, Harry Truman, a beleaguered public decision-maker if ever there was one, always kept his well-thumbed Plutarch handy, for quick reference. He fairly claimed that between them, Plutarch and Shakespeare (especially in the sonnets) had already answered most political dilemmas of any age.

John Aubrey's *Brief Lives* reflected a less lofty but quite durable purpose. He wanted only to preserve facts and anecdotes about noteworthy people of his time, but thanks to his vigorous language and contagious enthusiasm, he has outlived most of his subjects by three hundred years. If I were to play Charles Lamb's literary game and choose a writer to bring back from the past for an evening's conversation, I would pick John Aubrey over John Milton. Today Aubrey's eccentric, magpie voice is still arresting, and his purpose still drives much writing and reading of biography—to feed the common, primitive hunger for news about uncommon people. His lesser heirs churn out their books about the somewhat wicked, the very rich, and the merely recognizable, because they are there and because we are endlessly nosey, always drawn by life's variety and by whatever is secret. That such books keep appearing, and that we keep reading them, testifies to the lasting power of Aubrey's impulse.

A century after Aubrey, in the *Life of Samuel Johnson*, James Boswell created yet another kind of biography, the full record of a life in all its grandeur and foibles. It was as if one of Plutarch's worthies had walked off his pedestal and sat down in one's parlor. Boswell, life's eternal apprentice, had spent all his years stalking great men, and in this book he finally captured one completely. The great man uttered wisdom and pompous trivialities; he blinked, snuffled and endeared himself. Before Boswell, no one had written a life so vivid, fond and richly detailed; since then, no one has written one that is more so. One of the few life stories to match it is Boswell's other masterpiece, the drama he created about himself in his secret journals.

Then almost a century ago, in *Eminent Victorians,* Lytton Strachey stood Plutarch's purpose on its head. He indicted an era by sketching its heroes in acid. The lives he presented were not exemplary but contemptible; they revealed the rot and hypocrisy Strachey saw in his parents' generation. And Strachey stood Boswell's method on end, making his portraits not big and complex but small and consistent; the details all conspire to trumpet his theme of moral humbuggery. His knowing air and confiding irony beckon readers to join him in irreverence, and finally in scorn. He became biography's great master of dissuasion.

While Strachey was becoming an eminent anti-Victorian, yet another kind of biography emerged. It was not about moral exemplars or failures, the famous or infamous; it was not even about people. Some of the genre's early triumphs were about insects. Readers who usually lacked any interest in science were surprised to find themselves engrossed in such books as Jean Henri Fabre's *The Life and Love of the*

Insect (as one translator entitled his *Souvenirs entomologiques*) and Maurice Maeterlinck's *The Life of the Bee*. These books remained standard literary fare for decades. It was not new for naturalists to write about ants and bees, but it was new for a general audience to read about them in detail. They read both for pleasure and out of newfound scientific curiosity.

It helped that Fabre was both a great naturalist and a fine writer, and that Maeterlinck, a Nobelist in literature, could infuse tales of termites with lush prose and melodrama. Still, this does not fully explain why people who once cared only to swat insects now enjoyed whole books about them. Perhaps it was partly because a dazzlement of technology was changing people's lives so much and so quickly, and a public now intrigued by science could share naturalists' world more easily than that of chemists or engineers. Certainly Lyell and Darwin had changed people's vision of the natural world, placing humanity not above it but within it. I also suspect that as urbanization sped on at a dizzy clip, more readers were lifelong city dwellers, with a romantic, sometimes sentimental curiosity about nature, to which they felt like occasional visitors. Whatever the reasons, people usually more at home with poetry and novels now read avidly about nature, even its smallest visible members. Such interest would be even greater in their grandchildren's day, when saving whales and woodlands became a moral spectator sport.

Fabre and Maeterlinck were followed by other biographers of nonhumans. Some were inspired by the new sciences of ecology (organisms' relationships with their environments) and ethology (animal behavior in natural settings). Two of the best, Konrad Lorenz and Niko Tinbergen, were themselves distinguished researchers. Lorenz gave general readers his wonderful accounts of wolves, geese and jackdaws, and Tinbergen introduced them to his great studies of gulls. They did not win readers cheaply, by sentimentalizing otters and dolphins; they showed that species without big eyes and playful tempers can intrigue a curious observer and reward a patient eye. You don't have to be a specialist to become rapt along with Tinbergen as he observes and finally translates the black-headed gull's language of nods, cries, turns and bows. Nor to savor with Lorenz the moment when he squats, waddles and is imprinted by goslings as their mother.

These books are not intriguing just for what they tell about bees and sticklebacks; they also reveal how much we share with them. At first glance, many people see the animals' behavior as burlesques of human passions. With time and attention, they see that the opposite is true: our passions echo our evolutionary ancestors. However much we depart from the language of wolves, gulls and goslings, we still share with them a basic biogrammar. In these creatures we see the roots of human rage and tenderness, our need to be social and to communicate. Research is now revealing how the brain, hormones and neurotransmitters mediate such ancient strivings, and it suggests new ways to cope with stress, physical illness, disorders of mood and behavior, and social conflicts. The scientific world recognized how much ethology tells about human life and health in 1973, when Lorenz, Tinbergen and Karl von Frisch (who decoded the dance of the honeybee) were awarded the Nobel Prize in medicine.*

So life stories of animals and insects, like biographies of people, have a purpose. They illuminate the community of life, showing what makes species different and what unites them. They show that by knowing any creature, one better understands the living web in which it, and we, are threads. This is equally true of men and microbes. A bacterium may lack the instant appeal of a kitten or a meerkat, but it, too, can evoke a sense of wonder and kinship. If, as Blake said, there is a miracle in a grain of sand, there is much to amaze us in even the tiniest, least ingratiating creature. The life of *Borrelia burgdorferi* proves that no life is too small to be compelling or too short to hold revelations.

To anyone who still thinks a microbe too humble a subject, lacking drama and scope, I point out that thoughtful writers have found good subjects even below germs on the scale of life, in body parts and inanimate objects. There are

---

* Exactly one hundred years earlier, in 1873, the urtext of such studies appeared, Darwin's *The Expression of the Emotions in Man and Animals*. I swore when I started this book that it would have no footnotes or at least very few. Still, I must take this space to urge that anyone unfamiliar with Darwin's founding classic of human ethology have the pleasure of discovering it.

good books about salt, steel and the human hand. Reginald Reynolds's histories of beards and of toilets are learned and witty. George Stewart created engaging book-length biographies of a storm, a fire, a road and a rock.

The last word on biographies of nonhumans belongs to the late Professor Tinbergen, whom I had the privilege of knowing, and whom I found to be as sunny, inquisitive and charming as his books. At the close of *The World of the Herring Gull,* he said that even if decoding gulls' cries and head-tosses had no utility, he still would have watched them for years on Holland's windswept dunes, because "blood is thicker than water."

THREE

A Brief Aside Touching the Erotic Flea

A POINT ARISES THAT CALLS for either a very long foot-note or something kinder to readers, a very short chapter. I have hardly mentioned that most compelling of life stories, the firsthand account. I can imagine an ideal world in which one understood a germ's experience by reading its autobiography. Lest this seem entirely an exercise in flippancy, I should point out that I am not the first person to imagine tales told by lesser life-forms. If you went by book titles alone, you would think that at least one had already been written, by a flea.

The autobiography, like the biography, has a rich history and many purposes. Self-told tales range from religious testimony (Augustine) to various mixtures of confession and bragging (Casanova and Rousseau, the latter being perhaps the bigger braggart). In John Aubrey's day, cheap printing and a new literate class helped to create a growing market for memoirs, and they appeared in a flood. A new subgenre arose, the *faux* memoir, now widely malpracticed under the labels "fictionalized biography" and "nonfiction fiction." The best of the type is still one of the first, Defoe's *Journal of the Plague Year.* It purports to be a firsthand account of the appalling bubonic plague that struck London in 1665, but that year Defoe was only five years old. His "journal" was actually distilled from printed sources and older folks' reminiscences. That does not keep epidemiologists from calling it the best eyewitness account ever written of a plague. In its pages, said historian J. H. Plumb, "a metropolis dies before our eyes; the streets empty; grass grows where life reigned." It may also be the best memoir ever written by a witness who wasn't really there.

In Defoe's time, many people who retailed their life stories were gifted, adventurers or witnesses to amazing events. Today such distinctions are not needed, which is a mixed blessing. It is one thing when survivors of the Great Depression and Omaha Beach offer their "oral history." It is another when people with no dramatic gift fill journalism and daytime television with witless misdemeanors. Reveling in their four minutes of fame, they tell the world about vendettas by pets, lust for in-laws, and love-hate affairs with pasta or shopping. I, for one, would rather see autobiography, like biography, extended thoughtfully down the life chain, from ordinary people to mammals, birds and insects. I imagine how intriguing it would be to invert Defoe's classic and see the Great Plague of London through the eyes of the microbe that caused it, the flea that carried the germ, or the rat that bore the flea. In an apparent step in this direction, an anonymous Victorian wrote a book called *The Autobiography of a Flea.*

The tale is both more and less than its title suggests. It says little about fleas but a lot about luscious female bodies to which they might cling. Scholars have traced the conceit of the erotic flea back before the Victorians. In Marlowe's *Doctor Faustus,* the Clown pleads with Wagner to turn him into "a pretty frisking flea, that I may be here and there and everywhere. O, I'll tickle the pretty wenches' plackets." Later in the play, Pride says, "I am like to Ovid's flea; I can creep into every corner of a wench; sometimes, like a periwig, I sit upon her brow; or like a fan of feathers, I kiss her lips; indeed I do—what do I not?"

"Ovid's flea" is not Marlowe's invention but the title of a bawdy tale he and his audience knew well. It was already ancient, having appeared in many versions and many languages since the twelfth century and perhaps earlier. Why the lucky bug was called Ovid's flea is lost to history; there is, in fact, no amorous flea in Ovid's surviving works. However, in one of his elegies a man does savor the fantasy of being a ring he sent to his beloved, especially when she wears it in her bath. Perhaps, says some underemployed scholar, a later writer made the leap from an errant ring to a wandering flea.

In any case, from medieval jokesters through Marlowe, John Donne and Wilhelm Busch, writers and artists have reused the image of a flea touring a nubile girl and doing whatever fleas do instead of lubriciously wringing their hands. Victorian pornography gave the idea some elaborate twists, but its sly appeal seems to have faded in a day when girls wear garter belts as outerwear and S-M gear as accessories.

Thus the self-proclaimed autobiography of a flea does not match its promise. Neither, I am sorry to say, does a recent paperback book with a cover line announcing it to be "The Biography of a Bacillus." It is actually a reprint of Hans Zinsser's *Rats, Lice, and History,* first published in 1934. Zinsser was a cultured and supple (if sometimes prolix) writer and a world-famous bacteriologist. As he explained in the subtitle, his book is a "Study in Biography which . . . Deals With the Life History of Typhus Fever." *Rats, Lice, and History* was immensely popular when it appeared, and it never fell into neglect, for it is excellent medical history. Someone did it a disservice by slapping on the cover, as if it were the author's subtitle, "The Biography of a Bacillus." If Zinsser were alive, he would groan. As anyone knows after a week in Bacteriology 101, the tiny, round typhus germ resembles a bacillus, a larger rodlike microbe, no more than a mouse resembles a chimp. Zinsser would surely strike that line from the cover, insisting that he wrote a biography of typhus, not of a bacillus or of any other germ.

So much for autobiographies of fleas, bacilli and other low links in the Great Chain of Being. As much as I like the idea of a firsthand life of *Borrelia burgdorferi,* I am no Defoe, I cannot invent a convincing autobiography of a germ. Attempting the biography is challenge enough.

FOUR

Why Bb in Particular

MANY MICROBES COULD SERVE a biographer's purpose, but I have chosen *Borrelia burgdorferi,* or Bb as I will call it for short. I have no particular passion for this germ, though I do admire its elegant spiral form and the great adaptability it packs into a very small genome. Some other bacteria, such as the tuberculosis bacillus and the ubiquitous food poisoner *Escherichia coli,* affect many more people and do far more harm. But Bb is one of the most interesting faces, so to speak, in a large crowd; it offers a biographer many traits rarely found together in one microorganism.

First, Bb illustrates a wide range of facts about bacterial life. It also shows how a species can coexist with a vast and varied biological community, an ability Bb has in common with other widespread, adaptable organisms, such as humans. Bb and its relatives reside in hundreds of species, from ticks to reptiles, birds and mammals. Despite this ubiquity, its relationships with its hosts are so specialized that before Bb even enters a human it must first reside in two or even three quite different creatures, in a particular order. Thus Bb's life presents both a portrait in adaptation and a crowded ecological intersection at which one can watch myriad species meet and change each other's lives.

Bb is interesting for yet other reasons. It is affected by many aspects of its environment, such as the flora, climate and changes wrought by human activity. More than most germs, Bb responds to hardships and opportunities created by deforestation, suburbanization, global warming and cooling, and people's shifting patterns of work and play. Although humans and Bb are two of the least similar organisms on earth, their lives and fortunes are increasingly entwined—with us changing Bb's life at least as much as it changes ours.

There is yet another reason for choosing Bb. Much that is written about it is motivated by concern over Lyme disease. This literature leaves an impression that Bb was invented recently, by an irritable deity with no motive but giving us pain and anxiety. Yet Bb's ancestors and ours have coexisted for millions of years, and for most of that time they rarely met. Then in the last ten millennia, as people created farms and towns, they had a growing impact on Bb and raised the odds of meeting it. They built villages, felled forests, cultivated land and then abandoned or reused it; this created the scrub and second-growth forests that favor deer, deer ticks and Bb. These changes sometimes went so far that instead of helping Bb and enhancing its habitat, they threatened its survival; in some places they almost wiped it out. But the net effect, from the dawn of village life until a century ago, was that Bb and humans met a bit more often.

Then a century ago, wealthy nations began to repent having misused their environments; they started restoring woodlands and preserving wildlife. As we shall see, that was a boon to Bb and its animal hosts. Soon the germ was thriving again, even where it had almost been extinct. At the same time, suburbs mushroomed around growing cities and towns; people changed their lifestyles to feel closer to nature at home and at play. These were the events that finally put people in constant contact with flourishing populations of *B. burgdorferi*. A new balance evolved among the microbe and its habitat, animal hosts and new human neighbors. If this were a unique biological accident, it would deserve just a paragraph in biomedical history. But similar things happened to other microbes, through the same sort of technological, cultural and environmental changes. Bb's altered ecosystem, its increased contact with humans, and the resulting surge of Lyme disease offer a paradigm for the emergence of many other apparently new diseases.

There is yet another reason to choose Bb. Despite increased contact with people, it would have gone undiscovered if, like most microbes, it did not harm humans or domesticated animals. But inside people and their dogs, cats, cattle and horses, Bb sets off reactions that result in discomfort and sickness. Soon after this syndrome was discovered in Lyme, Connecticut, it was found worldwide, from

England and Germany to China and Australia. Now it is increasing in frequency and range, to become a global epidemic. And everywhere, as we are beginning to understand, the emergence of Bb and Lyme disease is a barometer of certain social and environmental changes.

Scientists hastened to study Bb and Lyme disease; in one human generation, the germ and the illness became subjects of a vast literature. One of its most impressive chapters is the mapping of the entire *Borrelia burgdorferi* genome, a task completed in 1997. Still, most writing about Bb has been done from a solely human perspective. By looking at the germ's life from its own point of view, one can better see the interplay of offense and defense between the microbe and its hosts, a contest that offers one of nature's most intriguing small-scale spectacles. Unlike many germs that enter people, Bb does not merely deliver a slap, fight a duel and quit the field; it conducts a lingering campaign that can range over the entire body and last for years or decades. The details of the encounter cast light on the nature of infection and resistance, illness and well-being.

For me, Bb has an added attraction as a subject, the irony that it ever became a common human pathogen. Lyme disease is not epidemic because people raped the land, massacred other species or despoiled their biota. Rather, it happened because people loved nature and became nostalgic for unspoiled settings; they tried to heal land they had exploited and ease urban crowding and blight. To a cynic, the emergence of Lyme disease confirms the adage that no good deed goes unpunished. To a naturalist, it is proof that an ecosystem's complexity almost always exceeds our power to grasp it. That is a truth we would do well to recognize before we repeat the mistake with deadlier microbes.

FIVE

Apologia Pro Vita Sua: In Defense of Germs

MANY PEOPLE FIND the very thought of germs disgusting. They make the word a synonym for human lowlife. Jimmy Breslin did so quite memorably when, after years of reporting on politics, he decided to run for public office himself. In his first and last race, he campaigned in New York State with Norman Mailer, on a platform of separate statehood for New York City. It was not a slate designed to draw professional politicians' heavy artillery. But even this mild inoculation taught Breslin that in politics, as in love and the stock market, one's cynicism may never be too great. The daily grime of campaigning proved even fouler than he had thought as an observer. The day after the election, Breslin made an audience howl with laughter when he blurted out in dismay, "My God! For months I've been surrounded by germs!"

So goes the common view of germs: they rival politics in squalor. But in reality, politics is dirty, germs are not. True, some microbes cause disease, and some thrive in stench, decay or squalor, but many more germs are harmless, and some are helpful or even essential to our survival. Strange to say, the popular association of microbes with dirt owes something to a Victorian movement bent on proving their innocence.

The idea that minute, invisible creatures cause disease goes back 2,400 years, to Democritus. Like his belief that all matter consists of invisible atoms, it took a very long time to gain favor. Time and again, from Roman days through the seventeenth century, the idea was revived or reinvented; time and again it was ridiculed and dismissed as fanciful. In the eighteenth century, it gained limited support, but by the 1820s, when deadly waves of cholera came sweeping out of Bengal and around the world, it was again in disrepute, especially among doctors. As cholera approached Europe, citizens there clamored for quarantines; their experience with such diseases as smallpox and syphilis told them that many illnesses somehow traveled from person to person. In many places, doctors and civil officials complied, reviving old regulations from the days of bubonic plague. However, they did so reluctantly, to keep foolish mobs from rioting. Among themselves, most experts laughed at the notion that creatures too small to be seen could fell armies and turn cities into cemeteries. It was like claiming that fleas murdered elephants. The few who did say that germs caused cholera were called cranks, clinging to the intellectual relic of an ignorant age.

Disease, most experts knew, was caused by miasmas, noxious vapors that arose from putrefying organic stuff. Microscopes did reveal tiny organisms in decaying matter and even inside living animals and people, but obviously that was because heat or decay had created them by spontaneous generation. Since some such creatures were present even in healthy people, it was absurd to say they caused sickness. Wise heads agreed that the villain was dirt. Where there was dirt, there was decay, and decay meant stench, the evil air that carried epidemic disease from victim to victim. (In Italy the phrase "bad air," *mala aria,* became synonymous with a devastating disease that thrived there in odorous marshes.) Even William Farr, the brilliant founder of epidemiology, warned that cholera scourged London because a deadly miasma prowled forth from its cesspools and sewers like a mad dog. In those days of headlong urban growth, Farr seemed to have vivid evidence on his side. London, like every European capital from Stockholm to Venice, was choking on its own wastes. The Thames was both London's sewer and its water supply; you could smell the river a mile away, and in some places you could almost walk on it.

Urban stench bolstered England's sanitary reform movement, which claimed that public and personal cleanliness, by ending decay and miasma, could prevent more harrowings by cholera. At mid-century, after two epidemics that probably compounded cholera with typhoid fever and other waterborne diseases, England loosed its reformers upon the land. Led by zealous Edwin Chadwick, they started cleaning up London's slums, air and water. They collected refuse, tore down hovels, built sewers and pulled dead dogs from the water supply. It was the beginning of something we now take for granted, massive government intervention to protect public health.

When cholera struck England for the third time, in 1866, there were fewer infections and deaths. Chadwick's legions said the improvement confirmed their theory that dirt caused disease; soon they were being imitated all over the world, with the creation of hygiene regulations, sewer systems and water filtration. What the reformers actually proved was that good results can flow from flawed theories; some kinds of dirt might invite and spread cholera, but they were not the cause. Therefore sanitation could reduce the epidemic but not eliminate it.

One London doctor, John Snow, praised sanitary measures but not sanitary theory. He had mapped cholera outbreaks in London and linked clusters of cases to specific sources of polluted water. His experiment with the Broad Street pump is epidemiology's equivalent of Newton's apple: by removing the handle from a stricken neighborhood's pump, he cut off the supply of contaminated water and ended cholera there. This, he said, proved cholera was caused not by dirt or miasma but by something in the water, probably something minute and living that could then be passed from person to person. Sanitarians called this notion foolish and dangerous. Evidence soon grew in Snow's favor, but for decades many people sided with Chadwick, who stubbornly denied that any disease could be caused by tiny, hypothetical hobgoblins.

By the 1880s, Louis Pasteur and others had proven once and for all that spontaneous generation was a myth; all cells rise from cells, all life from previous life. Pasteur and Robert Koch had also shown that specific microbes cause specific diseases. Koch pursued the cholera germ and finally, in 1883, during an epidemic in Cairo, he isolated the shimmying, comma-shaped bacterium *Vibrio cholerae*. It was just one of many triumphs in that early, heroic age of bacteriology. Almost every year someone found the cause, cure or prevention of another infectious disease. Yet the equation of dirt with disease did not die; rather, dirt and disease both became associated with microbes. This was not only because sanitation had succeeded in reducing some epidemics. There persisted an old magical association of epidemics with physical and moral filth.

That link was clear in reformists' fervent pursuit of spiritual as well as earthly pollution, which they saw as two sides of one coin. Their campaigns against cholera and typhoid were moralistic and sometimes downright paranoid. People in terror have always needed to blame someone for epidemics, to have lightning rods for their guilt and rage. When plague strikes, they wonder who inflicts such suffering on them, and what sin made them deserve it; then they project the sin onto others and point an accusing finger—at vengeful deities, the pitiless rich, the feckless poor, putrescence, poison or sin. Early Christians blamed pagan lust and riot. In the Middle Ages, mobs blamed the Black Death on the planets, lepers, Gypsies and especially Jews, who were burned alive in the tens of thousands. In the early twentieth century, native-born Americans blamed a new terror, epidemic polio, on poor immigrants from southern and eastern Europe. More recently, AIDS has been called heaven's rage at homosexuals, the devil's way with junkies, an invention of the Pentagon, and a CIA tool of racial germ warfare.

During cholera epidemics of the nineteenth century, people conjured up the usual villains. Many of the poor thought cholera was poisoning by the rich to wipe out restless peasants and demanding workers; others said doctors were spreading toxins to get corpses to dissect in medical schools. In some places, they beat government officials and stoned doctors, sometimes to death. The comfortable classes thought poor people had invited cholera into society by being unemployed, dirty and drunk. Thus Victorians, with their tendency to moralize every itch of body and soul, linked moral, social and physical debility. Many backed sanitary reform, which blamed people for being sick, and ridiculed germ theory, which was morally neutral. Disease, they said, was like dirt, stink and indolence; it could be ended by moral purpose and strength of will. From pulpits and editorial pages they warned that the habits of the poor bred miasmas that wafted disease to their betters. If the poor would stay out of literal and figurative gutters, corpses would not be piling up in good homes as well as bad.

Such ideas were held by otherwise decent, intelligent people. Scientists and clerics agreed that the higher rate of cholera among the ill-fed proved their lack of industry, self-control and innate fitness. England was not alone in confusing medicine and morals; these beliefs were common all over Europe. There was less stress on class in the United States, but here, too, as wagons rattled over New York's cobblestones to collect the dead, fighting cholera meant church bells, public prayer and calls for sobriety, contrition and baths. Since the baths increased people's exposure to vibrio-laden water, they might have been safer staying dirty.

By the twentieth century, germ theory had triumphed in most people's minds, but many Americans still confused physical and moral laundering. Hygiene became a social ideal, entwined with fear of the poor and the newly arrived. When settlement houses and visiting nurses taught immigrants to use soap and comb out nits, they were protecting more than sensitive noses. They feared that Irish, Italian, Jewish and Slavic immigrants were polluting a once sturdy Anglo-Saxon nation with typhoid and tuberculosis. The poor and foreign-born were also thought to spread moral toxins. A cry went up that immigrant prostitutes tainted WASP clients with syphilis; those stricken roués became traitors to their race by passing the disease to their wives and making them sterile, and by siring congenital idiots with saber shins and notched teeth. Syphilitic sterility, they said, was lowering the Anglo-Saxon birthrate, while lesser breeds multiplied like rodents. Every aspect of public and private life seemed vulnerable. The *New York Times*, among other sources of edification and anxiety, warned that Americans might now be traveling at their peril in trolleys, trains and ships steered by syphilitic lunatics.

Visible and invisible filth were thought to cause not only polio and syphilis but the "Spanish flu," which near the end of World War I killed more than a half-million Americans and twenty million people or more worldwide. Several decades would pass before the invention of the electron microscope, which revealed the flu and polio viruses; meanwhile people flailed in vain at these epidemics by attacking every source of muck and germs they could think of. They washed buses, scrubbed public telephones, scoured drinking fountains and wore surgical masks in theaters and restaurants. They were advised that after leaving stricken (poor and immigrant) neighborhoods, they should change their clothes, bathe, shampoo, gargle and spritz disinfectant up their noses.

None of this stopped influenza, and there was always evidence that polio preferred the clean and affluent, striking suburbs harder than slums—only decades later did researchers see that dirty, crowded cities spared children polio's worst ravages by exposing them early to its mild forms. In the early twentieth century, nothing could stop the panic and scrubbing. Even many scientists remained convinced that dirt nourished germs, germs bred disease, and nothing human was as dirty as poor foreigners. It may require another book to explain why Americans of African and Asian descent did not become bigger targets of para-noia; perhaps their firmer exclusion from mainstream America made them seem less likely suspects.

The war against moral filth was as futile as the scrubbing of phones and fountains. A hundred years ago, syphilis filled more hospital beds than any contagious disease except tuberculosis. Untreated it could lead to blindness, madness, paralysis and death; this fate, which struck Maupassant, Nietzsche and Sir Randolph Churchill, haunted the Victorian imagination.* The discovery in 1905 of the syphilis microbe, a pale corkscrew related to *B. burgdorferi,* seemed to promise the end of a nightmare. The next year, August von Wassermann devised a test to detect the germ, and in 1909 Paul Ehrlich invented Salvarsan to combat it. This was the first chemical "magic bullet" targeted against a specific microbe, the precursor of sulfonamides and antibiotics. Though not fully effective or safe, it was far better than anything before.

The stage was set in America for an all-out campaign against the deadliest of "social diseases."† By World War I,

---

* Or more properly the pre-antibiotic imagination. See Isaac Babel's unforgettable short story "Guy de Maupassant," written in the 1920s.
† The phrase "social disease," now quaint, was still current in my youth as a half-humorous euphemism. It was coined in all seriousness in the early twentieth century by Prince Morrow, a physician and reformer who taught at New York University. Like many people then, Morrow thought syphilis and gonorrhea were spread chiefly by prostitution, which polite people called "the social evil"—hence social diseases. In

marriage was impossible in most states without a Wassermann test, and our military chiefs had resolved that American innocents, many fresh from farms, would not be infected in big-city brothels or Gallic nests of erotic cunning. The army shut down America's red-light districts, and in France enlisted men were encouraged to play lots of volleyball. Posters depicted syphilis as a grinning skeleton that crooked a beckoning finger at youth.

All of which accomplished next to nothing. Hookers moved from brothels to automobiles, and volleyball proved resistible. Reformists' moralizing and fear tactics repelled some people; they merely made others laugh. Many were scared away from testing and treatment. Until the advent of penicillin and smarter public-health campaigns, syphilis went on causing blindness and paralysis. A half-century later, many AIDS-prevention programs proved the persistence of human shortsightedness by repeating the mistakes of the early war against syphilis, from moralistic rants to calls for premarital blood tests.

I do not revisit these follies about germs, dirt and sin to enjoy an easy laugh at people who lacked the foresight to be born a century later. Rather I want to emphasize that such

---

those days, the New York Times would not print the word syphilis; it reported that Ehrlich had won a Nobel Prize for inventing a cure for "a blood disease." I learned of Morrow's having invented the phrase social disease when I was a doctoral student at New York University, where no one seemed to know or care that its once illustrious author had been one of their own.

myths preceded and survived them. Today as in the past, many people think of disease as a punishment for sin or swinishness. The microbes that cause it are still seen as malevolent; from daily speech to cartoons, we depict them as revolting, buglike creatures with many legs, bulging eyes and avid leers, poised to sink their fangs in us. We still tell children that dirt and germs are synonymous, and that both will make them sick.

It is no surprise, then, that microbes arouse little sympathy, sometimes not even the naturalist's usual neutrality. In recent debates about incinerating the world's last known vials of smallpox virus, there has been much talk about the great military and medical implications, but little about the fact that this would be the first time humans deliberately made another species extinct. The prospect inspires no angry placards, no demonstrations before television cameras. I do not urge a Save the Virus movement; I merely point out that our mental bookkeeping tends to assign otters and owls to one page in the book of life, microorganisms to another. If some unenchanting little mollusc were thus threatened, a passionate crowd would march to save it, if only because it might play an unknown role in nature's economy. Yet most people would laugh at the idea of protecting a germ. And that is what many microbes merit.

The more one learns about microorganisms, the more one appreciates them. Most of us know that they turn fruit into wine and milk into cheese, but these are minor accomplishments. Microbes are indispensable to all other forms of life.

They help maintain the earth's atmosphere, enrich the soil, nourish plants, aid cows in digesting grass, give sea creatures luminescence, and manufacture vitamins in the human intestine. Most important, they destroy organic debris through decomposition. If they did not, every plant and animal that died would remain where it fell, and the surface of the earth would vanish under heaps of corpses. Very soon all life would cease. The decay once thought to cause disease is the means by which microbes reduce life to its elements, so it can re-create itself.

Little more than a century has passed since the dawn of bacteriology, and half a century since antibiotics appeared. Since then, the ability to cure or prevent many diseases has provoked fantasies of life without infection, a glistening, sterile future from which microbes had vanished. But it was always absurd to think that we could or should keep germs out of our lives. Microorganisms—bacteria, viruses, protozoa, yeasts and molds—make up 80 to 90 percent of the earth's biomass, and we meet them by the millions every second. They fall on us from the air, enter us with water and food, are rubbed into our skin, and travel on everything we touch. We shed them in wastes, saliva and tears, by coughing, sloughing off skin and exhaling. Even if we could avoid most of them, there is no reason to, for they have no ill effects. Only a small minority can make us sick, and most of those die trying. In fact, thousands of kinds of germs are native to our bodies, five hundred to the mouth alone, and without many of them we would be ill.

So we owe it to ourselves to know some of the facts of microbial life. The tools of modern laboratories reveal that we have always walked in ignorance through an invisible world that hums with tumult. Microbes constantly attract, repel, invade and consume each other. At the genetic and molecular levels, they carry on a blizzard of communication with each other and their hosts. They pass genes back and forth, transforming themselves to adapt and readapt to human activity. We enhance the process by carrying them worldwide and transplanting them into new biological communities. Now we are genetically redesigning microbes and putting them to work, from bacteria that eat oil spills to viruses that deliver helpful genes to ailing human bodies.

Obviously germs do not deserve disgust or reflexive fear. Some must be treated warily and others thanked; all should make us attentive. The variety and ingenuity of their survival mechanisms is awesome. It is natural that when speaking of microbes that make us sick, we fall into the rhetoric of warfare, as if bacteria, like Farr's cholera miasma, stalked us in rabid rage. But that is just a metaphor for what it feels like to be on the other end of a natural balance. Pathogens lack malice; they are just trying to survive, and sometimes they must do so at other creatures' expense. The same could be said of humans.

SIX

In Some Warm Little Pond

THIS IS THE CHAPTER many biographies put first, with a title like "Honest Yeomen" or "A Daughter of Princes." It describes the protagonist's earliest ancestors and the source of the family name. If the historical record is thin, the result is a sketchy genealogy. If it is rich, there may be glory or scandal—perhaps a title bestowed on some thug who pleased his lord by gutting unbelievers or cropping peasants' ears. Sometimes such stories make one thankful that today official honors are more often bestowed on tenors and comedians than on defenders of the faith and protectors of landlords.

The family tree of *Borrelia burgdorferi,* however, lacks princes and tenors, and the germ's ancestors may have lived in a quiet, modest place. So Darwin would have said, who speculated that life began "in some warm little pond." This placid picture of life's birthplace dominated scientific thought for a century, and a sun-drenched pond or tidal pool may indeed have spawned Bb's forebears. However, lovers of ancestral drama will be glad to know that research now suggests more tempestuous cradles of life—hot volcanic vents on the ocean floor or near-boiling mineral springs. If these recent theories are correct, the first living creatures, ancestral to Bb and all other organisms, emerged in surroundings that today would kill almost anything on earth.

There are many theories of life's origin, some convoluted and quite speculative; not one is universally accepted. Some people say it beggars logic that anything as complex as the simplest cell developed by chance through the chemical processes seen in nature today. They assert that life, or at least some of its major components, rode to earth from outer space on meteorites. (A few also claim that the influenza virus reached our planet as an extraterrestrial hitchhiker; why that virus alone had a free ride is not clear.) It is true that for life to have risen from a random wedding of elements seems almost miraculous. However, it seems equally miraculous that anything living could survive a voyage through space, the entry into earth's atmosphere, and the shock of impact. Some proponents of the life-from-space (panspermia) theory make ingenious cases, but they seem to solve the question of origins by conveniently exporting it to another galaxy. I myself do not see why life or its ingredients should arise there any more easily, so I will take the matter no further.

When biologists speak of life, they usually mean an organized, reactive unit that reproduces through the action of DNA. DNA, of course, is the stuff of genes, a double-stranded helix that acts as a template for creating the proteins from which all living things are built. The origin of DNA has therefore been seen as synonymous with the origin of life. Long before the discovery of proteins, genes and DNA, Darwin guessed that life began when solar energy stimulated the chemical components of life in shallow waters. In retrospect, this pastoral vision of a warm pond, with its hushed and tranquil air, may reflect Victorian reverence for nature as much as it does real history. Still, it held sway in scientific and popular thought for a century. It came up for revision in the 1950s, when a famous experiment by Miller and Urey showed that lightning or ultraviolet radiation could have turned compounds in the primordial oceans into amino acids, the building blocks of proteins. In many scientists' minds, Darwin's quiet pond was replaced by a stormy, lightning-struck sea.

The new science of molecular biology further elaborated life's origin by mapping the structure of DNA. The picture was complicated by the discovery that RNA, a simpler, single-stranded molecule that helps DNA to replicate, sometimes copies itself. This explained another discovery, the existence of viruses that contain no DNA at all, only RNA. Now it seemed conceivable that very simple organisms based on RNA alone preceded those with DNA. Perhaps, some now say, life began at an even simpler level, guided not by RNA but merely by one or more of its constituents.

Another piece was added to the puzzle in the 1970s, with the discovery in seabed volcanic vents and steaming springs of primitive, bacteria-like organisms called archaea. They are probably the closest thing around now to the ultimate ancestors of *Borrelia burgdorferi,* and perhaps of everything else that lives.

As these ideas developed, researchers expanded the time frame in which life arose and Bb's ancestors appeared. Older scientists still remember when the earth was thought to be two or three billion years old, and life half that age. Now it seems that our planet was born five billion years ago. A billion years later, the oceans had stopped boiling, and primitive organisms appeared. Just how and where they did so is moot, but somewhere, in sunny pools or hot ocean vents, chemical reactions gave rise to a soup of organic compounds, and they combined and recombined until something took shape that could reproduce itself. Biological processes took over, and mutations created a growing variety of forms. These proto-life forms developed envelopes to contain an ocean-like microenvironment and became the first living cells.

The first one-celled creatures gave rise to three kingdoms, or major divisions, of life that still exist today, the archaea, prokarya and eukarya. Least complex are the archaea, which have no nuclei or organelles (distinct specialized regions). They live in extreme environments—hot, often salty, and lacking oxygen—that used to be thought too harsh for anything alive. There are archaea dwelling in seabed vents that endure a thousand times the pressure on the earth's surface. Others thrive in springs that feed geysers in Yellowstone Park, at temperatures just below boiling. More are still being discovered in the Great Salt Lake, the Dead Sea and even in smoldering coal heaps. They feed not on organic matter but on minerals such as sulphur and magnesium.

Bacteria evolved from archaea or from some common ancestor of them both. The world's most ancient fossils, more than 3.5 billion years old, are those of one-celled organisms resembling today's blue-green algae. Despite their name, these are not plants, as true algae are, but very simple bacteria. Bacteria, like archaea, have no nuclei, but there are important differences between them. Few bacteria can survive extreme environments, as archaea do; many thrive at moderate temperatures and in shallow water—blue-green algae are just the sort of pond scum that inspired Darwin's image of early life. But the most important difference from archaea is that some bacteria perform photosynthesis, using sunlight to create energy and releasing oxygen as a by-product. That was the development that opened the door to life as we know it.

Oxygen is the fuel of animal life, and in the days of the first bacteria, this planet's atmosphere held almost none of it. When some bacteria started releasing oxygen, they created a new ecological niche; and as happens when any environment changes, some species rushed in and adapted to take advantage of the situation. New bacteria evolved that burned oxygen, which proved to be a far more efficient kind of metabolism than anything before. When oxygen approached its present level in the air, 600 million years ago, it helped set off a storm of rapid evolutionary change that created a multitude of complex animal forms. Photosynthesis, oxygen and animal life are all, in a sense, bacteria's gifts to the world. And bacteria still have vital roles in recycling oxygen, nitrogen and other life-sustaining elements.

The most complex kingdom, the eukarya, appeared more than two billion years ago; from the first ones descended all present-day plants and animals, from one-celled varieties to orchids, opossums and human beings. They differ from bacteria in several ways. First, some bacteria depend on oxygen, but all eukarya must have it or die. Eukarya pack most of their genes into a central walled nucleus, and they have organelles such as mitochrondria, small bodies that control their energy production. Virtually all bacteria live on substances dissolved in their liquid environment; eukarya can engulf and digest large organic molecules or even entire one-celled organisms. Bacteria are relatively slow reactors; nutrients and chemical messengers travel through their protoplasm by diffusion. Eukarya, crisscrossed by networks of tiny tubules, shunt energy and enzymes quickly through their interiors.

Rich in energy and quick to respond to their environ-

ment, eukarya may seem to have an edge over bacteria.*
Indeed, they have kept evolving toward great complexity
and specialization. Some became highly mobile indepen-

---

* Some readers are passingly familiar with the microbial world. Those
who are not may find use for this very abbreviated guide to one-celled
life.

Microbes: Organisms too small to be seen by the naked eye. The term
includes all of the following except microbiologists.

Germs: Microbes, especially those causing disease.

Archaea: Very simple bacteria-like organisms that live in extreme
environments.

Bacteria: One-celled organisms without nuclei or organelles. When
many people say germs, they mean bacteria, though most bacteria
do not cause disease, and some other kinds of microbes do. Usually
included among bacteria are so-called blue-green algae, which are
harmless, and very tiny organisms called rickettsia and mycoplasmas,
which are not.

Prokarya: Cells without nuclei or organelles; this category includes
bacteria.

Eukarya: The cells of all plants and animals, whether microscopic or
large; they all have nuclei and organelles. Among them are protozoa,
fungi and true algae (one-celled plants).

Protozoa: One-celled animals. Some cause disease in larger creatures,
including man.

Parasites: Strictly speaking, all creatures that live in or upon others, at
the hosts' expense. The term includes viruses, many bacteria and some
protozoa, arthropods and worms. When microbiologists (see below)
speak of parasites and parasitology, they usually mean the invisible
ones, especially protozoa, but sometimes also larger ones such as lice,
mites and worms.

Viruses: A virus consists of genes in a protein coat. To function and
reproduce, it must inhabit a host cell and borrow its metabolism; the
host may be anything, a bacterium or tulip or human being. Viruses are

dents, the protozoa. Others huddled in groups and began to function in concert; this opened the evolutionary path to worms, insects, fish, reptiles, birds, mammals and, three million years ago, the first hominids. The cells of higher plants and animals became increasingly varied; in a human body they range from the threadlike neurons that carry nerve impulses to amoeba-like white blood cells that patrol the body and engulf microbial invaders.

Bacteria did not take the eukarya's vertical route to greater complexity and group life, but they hardly became obsolete. Instead they took a horizontal path to diversity, evolving into a multitude of forms adapted to very different environments. Billions of years after they first appeared, they continue to adapt and diversify. Anyone who has been colonized by Bb and other disease-causing bacteria knows how well they can contend with the "higher" cellular forms that compose the human body. Like sharks, ticks and turtles, they are so superbly designed that they have outlived many of the larger, more complicated life-forms that developed after them.

Some time in the distant past, probably hundreds of millions of years ago, Bb's bacterial ancestors emerged from the seas where they originated and began living in marshes, mud

---

minuscule compared to even the smallest bacteria and can be seen only with electron microscopes.

Microbiologists: Scientists who study subvisible organisms. Some specialize, calling themselves bacteriologists, virologists or parasitologists; the latter study protozoa and larger parasites such as worms.

flats, and finally on dry land. For many of them, this meant changing their shape, outer walls, feeding habits and metabolism. Some learned to survive periods of extreme dryness, heat or cold by forming spores and lying dormant for long periods. But bacteria are by nature aquatic; to survive, most of them need at least a microscopic film of moisture, from which they can absorb the water-soluble molecules that nourish them. Some discovered that the moisture of a living host's interior was more than sufficient.

As a group, bacteria are sturdy opportunists. They have managed to adapt to almost every environment on earth. They live in fresh and salt water, in deserts and arctic wastes, deep in the earth and high in the atmosphere. Some perform photosynthesis; others eke out a living by digesting organic materials, gases, wood or even minerals or asphalt. Some cannot live in oxygen; others burn so much of it that their presence saves metal pipes from corroding. And many have evolved along with the animal kingdom by living inside it.

Inside a host, bacteria must cope with that creature's temperature, acidity and immune system. It is worth the effort, because there they are sheltered from the fluctuations of the outside environment and have a ready-made larder. Many bacteria specialize not only in certain hosts but in particular tissues. The cells of a termite's gut and human mouth are very different physical and chemical worlds, but bacterial species have learned to live in each one, and only that one. Most of them do so without endangering their own lives by provoking illness and thus mobilizing host defenses. Some very versatile bacteria, such as Bb, have adapted to a variety of hosts; this raises the odds that they will survive if any one of them becomes unavailable.

I have tried to briefly sketch Bb's time and place of origin, and the vast, varied microbial realm to which it belongs. Now, having located bacteria's region in the tree of life, it remains to locate the branch holding Bb's extended family, the spirochetes, and Bb's particular twig. Assigning organisms to twigs and branches is not an arbitrary labeling game but a demanding scientific specialty. Since a creature's name proclaims its nature, it must be as accurate as a chemical formula or a magical incantation. Otherwise the study of life becomes chaos, and the givers of names are sorcerer's apprentices. That was pretty much the state of things before Linnaeus, whose system of names was the first to create a coherent picture of life on earth.

SEVEN

Linnaeus's Tree

WHEN WE CALL SPIROCHETES A FAMILY, *Borrelia* a genus and *Borrelia burgdorferi* a species, it sounds as if such categories were part of nature. But nature holds no stone tablets declaring this is a family or that a species. Such categories are created by systematists (also called taxonomists), who decide which traits define a species and which species form a family. It is they who make the case for or against using such concepts as spirochete, mammal and hominid. If many of their conclusions seem obvious, even self-evident, it is because they are so familiar. Only when you try to look at nature without them do you grasp how much thought and observation went into creating them.

To give creatures suitable names, the systematist must first observe, observe, observe. Then he weighs the similarities and differences between organisms, to decide which traits should have priority. However, nothing tells him in advance whether to compare their outsides or insides, their size or shape, how they move or how they reproduce, their brains, toes, teeth or diets. Are the mosquito and bluebird close cousins because they both lay eggs and fly? Or are their differences more important? The bird has two legs, the mosquito six; the bird's skeleton is inside, the mosquito's outside; the bird eats solid food with a beak, the mosquito sucks up its liquid diet through a tube. When the systematist decides which of these features to give defining power, he cannot be hasty or arbitrary. Classifications endure only if they explain more than alternatives do, withstand endless reexamination, and accommodate new facts as they appear.

When scientists decide where to place an organism in their scheme of nature, they give it a two-part scientific name. This double-barreled Latin tag is not meant, as Ezra Pound once testily claimed, to conceal it from vulgar knowledge, but to state its address in the living community. Like most addresses, it reveals a great deal, so one does well to look it up. If the thing is a germ, you will consult *Bergey's Systematics,* which is to bacteria what the OED is to the English language and the Elias Sports Bureau to baseball statistics. It is not the whole truth or eternal truth, but for certain purposes it is the best truth available. *Bergey's* itself is quick to point out that species and genera, created by fallible humans, must sometimes be revised or replaced. With an owlish attempt at humor, *Bergey's* most recent editor (Bergey himself is long gone) reminds readers in boldface type that **"bacterial classifications are devised for microbiol-**

ogists, not for the entities being classified. Bacteria show little interest in the matter of their classification."

Well, not exactly a barrel of laughs, but the point is well taken, and as the boldface warning implies, even some scientists must be reminded of it. Most of us first learned biology from uninspired textbooks that presented nature as a fixed system of clearly defined types. We were sent on little expeditions to parks, bearing illustrated charts that showed us how to tell oaks from mimosas and sparrows from starlings; their forms seemed as clear and immutable as the planets. But later, if we studied nature more deeply, the texts and charts sometimes failed us. In biology lab, dissecting our first frog, we cursed as we searched for nerves and veins that weren't where the book said they were. Collecting flora in meadows or fauna in creeks, we saw that living things often violate the charts, confounding definitions and blurring the boundaries between species. Some variations on basic patterns, such as the four-leaf clover and the black swan, are obvious and easy to see, but many more are subtle, even invisible, involving diet, enzymes or immune defenses.

Evolutionary theory sees an advantage in such variety, for it raises the odds of adapting to environmental change. The propensity for mutation, and thus for variation, is especially strong in bacteria, and it operates in them at high speed; they can go through tens or hundreds of thousands of generations in one human lifetime. And since they have been diversifying at this breakneck pace for several billion years, they are much more varied than birds or mammals; Bb and the bacilli in sour milk differ from each other more than hummingbirds and eagles do. There are myriad bacterial types, most of them still undiscovered or undefined; furthermore, some species have dozens or hundreds of barely distinguishable subtypes. Trying to identify a given bacterium can evoke all the frustrating uncertainties of dissecting one's first frog.

It is no surprise, then, that systematists often debate and sometimes reject predecessors' work. What does surprise is that they still use a system more than two and a half centuries old. Swedish naturalist Carl Linnaeus devised it because the categories he had inherited were becoming a bad joke; they resembled not a family tree but a sprawling, untidy warren. Most of his colleagues still followed Aristotle in lumping newts and cats together as quadrupeds, although one laid eggs and the other bore live young. It was rather like lumping mosquitoes and bluebirds together because both have wings. Some naturalists even put different parts of one animal in disparate categories; the results, when fitting newly discovered beasts into the old system, now seem whimsical. For instance, in 1721 the Jesuit explorer Pierre de Charlevoix wrote a book about New France, as French territories in North America were then called, and because its economic mainstay was beaver furs, he took pains to describe this New World animal to his countrymen (I condense):

It is an amphibious quadruped which cannot live for any long time in the water. Its legs are short, particularly the forelegs, pretty much like those of the badger. The hind feet are quite different, being flat and furnished with a membrane between the toes. In respect of its tail, it is altogether a fish, having been juridically declared such by the faculty of medicine in Paris, in consequence of which declaration, the faculty of theology have decided that it might be lawfully eaten on meager days. Monsieur Leméry was mistaken in saying that this decision regarded only the hinder part of the beaver. It has been placed all of it in the same class with the mackerel.

Such was systematics in Linnaeus's day. It told Catholics that on meatless Fridays they could eat the entire beaver, not just the tail; otherwise it spread more murk than light. Linnaeus invented a lasting alternative. After studying medicine at the university of Lund, he turned to botany and devised a revolutionary classification of flowers, based not on their blossoms but their reproductive organs. A plant's stamens and pistils, he said, characterize it better than the "perfumed bridal bed" of its petals. His prose may have been overripe, but his thinking was crisp. Eventually Linnaeus classified more than ten thousand plants and animals according to a grand scheme that he developed over a quarter of a century. Later some of its details were changed or abandoned, but the basic system survived the revolutions of Darwinism and molecular biology, which swept so much else away. Many of its categories are still in use, so familiar that they no longer seem like intellectual lightning strokes. For instance, Linnaeus noted that only the hairy quadrupeds with warm blood have milk glands for feeding their young. This trait, he realized, told at least as much about them as their leg count, and to describe them he coined the word mammal.

The Linnaean system puts all living things—plants, animals, bacteria—in a hierarchy of categories, descending from kingdom to division, class, order, family, genus and species, with room below for subspecies, varieties and strains. Every organism receives a Latin name consisting of a noun (its genus) and an adjective (the species); each part of the name reflects an outstanding trait or the discoverer's name. *Borrelia burgdorferi* belongs to the kingdom Prokaryotae (bacteria), division Gracilicutes (bacteria with thin cell walls), class Scotobacteria (bacteria that do not use light for energy), order Spirochaetales (spiral bacteria), family Spirochaetaceae (spirochetes with particular cell characteristics), genus *Borrelia* (a group of tick-borne spirochetes discovered a century ago by bacteriologist Amédée Borrel), and species *burgdorferi* (for its discoverer, Willy Burgdorfer).

The Family Tree of *B. burgdorferi*

Kingdom Prokaryotae
Division Gracilicutes
Class Scotobacteria

Order Spirochaetales
Family Spirochaetaceae
Genus *Borrelia* (related genera are *Cristispira,*
Treponema and *Spirochaeta*)
Species *B. burgdorferi* (with twenty other species
of *Borrelia*)

In my early teens, when I became fascinated by biomedical science, I learned the bare bones of Linnaean nomen-
clature, wanting to understand the names of things. The
system's grandeur bowled me over; it gave every living thing
a name, and each name told a story. Latinate gabble became
enlightenment, and Linnaeus became one of my intellectual
heroes. (Anyone who mistakes this for early maturity has
forgotten that boys are natural systematists who may passionately collect and classify anything from baseball cards to
beetles.) Having started to gather and dissect specimens, I
could begin to imagine how much knowledge and imagination went into creating such an edifice. Linnaeus seemed to
me as bold as Darwin and as creative as Bach, and he still
does.

Incidentally, almost forty years later I got to sit under Linnaeus's own tree. I had gone to the medieval university town
of Lund to attend a month of sexology seminars; Linnaeus
was far from my mind. On the first day, during a tour of the
grounds, my guide pointed out a little brick summer house
built for the Danish kings who once ruled southern Sweden.
Next to it stood a broad, leafy old tree. "That," he said, "is

Linnaeus's tree." No, he assured me, this was not an elephantine academic joke about the Linneaean tree of life. Having studied in Lund, Linnaeus returned there to teach, and local tradition says he planted this tree and later liked to sit under it. During my stay in Lund I often sat and read in its shade, bemused by the odd circles life sometimes completes. It felt like sitting at Newton's desk or Bach's organ. I never checked out the story, since I so enjoyed believing it.

EIGHT

And Bb's Twig

LINNAEUS WOULD PROBABLY NOD in recognition at the sight of systematists scuffling in his intellectual garden, at odds about what defines certain microbes. Bacteria were first labeled according to the most obvious fact microscopes revealed, their shape; they were classified as rods (bacilli), spheres (cocci) or spirals (spirochetes). Then researchers learned more about how they function—where they live, at what temperatures, whether they are mobile, whether they cause diseases, how they are transmitted, the chemistry of their outer walls, and what they feed on. For instance, all bacteria need carbon, oxygen and nitrogen, but some also need manganese or lactose. Some produce methane gas, while others release sulfates that become sulphuric acid and can eat away metal or stone. Such facts became the basis of new classifications; later so did the knowledge gained from electron microscopes and molecular biology. Today the mapping of bacterial genomes promises a major overhaul of all bacterial categories.

That promise has made some people call the Linnaean tree an antique. Certainly it is giving ground to new systems. Besides the one rising from genetic profiles, there is another called cladistics, based on evolutionary relationships. Still more are likely to appear, based on comparisons at the molecular level. Yet the fact remains that when new genetic varieties ("genospecies") of *Borrelia burgdorferi* are discovered, they are christened as all species have been since 1758, with the double-barreled Latin names—*B. garinii, B. afzelii, B. japonica,* and so on. Epitaphs for the Linnaean system are still premature.

And despite all the new genetic and molecular knowledge, the first fact you are likely to read about Bb is still its corkscrew shape, which enables it to wriggle snakewise through watery environments. Like all flexible, coiled germs, it belongs to the order Spirochaetales, a fair-sized bacterial sub-branch in the evolutionary tree. It is anyone's guess when spirochetes first appeared, but it may have been hundreds of millions of years ago, in ancient seas. Some still live independently in oceans or rivers, but the majority have become parasites, feeding on hosts' fats, sugars and other ready-made nutrients. One species of parasitic spirochete inhabits molluscs, but most have developed specialized ways of life in land-dwelling hosts from ticks to mammals.

Inside their hosts, spirochetes have affinities for certain tissues, such as kidney cells or brain cells. One spirochete

dwells only in the gut of a single species of termite, another in the human intestine, and several in the crevices between people's teeth and gums. Oral spirochetes usually do not trouble us, though they can cause gingivitis if poor hygiene ignites a population explosion. This relatively benign relationship means that they are probably long-time residents. A microbe that is new to a host does not meet finely honed defenses, so it usually causes acute symptoms. In time, germ and host tend to form an *entente cordiale,* a balance of offense and defense; the result is low-grade illness, chronic illness, or none. So our rather peaceful coexistence with oral spirochetes implies that we acquired them long ago; we may even have inherited them from our primate ancestors.

There are three major groups of spirochetes that make humans sick, the genera *Leptospira, Treponema* and *Borrelia*. Each genus has distinctive traits, but there are some they all share. Consider first the *Leptospira* (slender spirals); they are thin, tightly coiled and hooked at the ends. One species, called *L. interrogans* because it vaguely resembles a question mark, makes people woefully ill. Its original home was mice and rats; it did not infect humans until they settled down as farmers. Then people lived fur by jowl with their herds and pets, and their granaries and garbage drew scavenging rodents; with rodents came their parasites. Thus a new intimacy with human settlements swept leptospires into a new infectious cycle linking people, rodents and domesticated animals.

L. interrogans favors hosts' livers, brains and especially their kidneys, causing symptoms from mild fever and jaundice to kidney failure. Infected mice, dogs and cows all shed the germ in their urine; it hardly survives in farmyard puddles and the streams they drain into; new hosts swallow it or pick it up through broken skin. Leptospirosis is uncommon now in prosperous nations, and it can be cured with antibiotics, but it remains an occasional hazard to American farmers, veterinarians and meat packers, and to vacationers swimming in water polluted by farm runoff. And as American soldiers learned in Vietnam, it can be caught by slogging through swamp or jungle. Leptospirosis is also called mud fever, harvest fever, swineherd's disease and pea picker's disease; like Bb, it usually enters people because of how they live and work. And like Bb, it became a human pathogen when we stumbled into its natural life cycle and unwittingly amplified it.

The other genus of worrisome spirochetes is the treponemes (turning threads). Unlike leptospires, they are very delicate and easily killed by heat or dryness; they must pass directly from host to host. A dozen or more species live harmlessly in various mammals, but *T. pallidum,* the pale treponeme, is one of the most interesting and intensively studied of all bacteria. It causes not one but four diseases, and in some ways its effects on hosts resemble those of Bb.

According to one theory, this germ's first host was an African primate, and twenty to thirty thousand years ago it spread to people as well. (Later the original host and/or the ancestral microbe must have died out, for today *T. pallidum* infects only humans.) In the tropics this spirochete caused pinta, a skin disease that children spread by casual body contact. About 10,000 years ago, a mutation made the germ more virulent, able to attack bones as well as the skin; the result was yaws, which like pinta still exists in hot, humid parts of Africa and Latin America. In time, *T. pallidum* spread to Neolithic villages in dry, cool climates, where people were fully clothed all year round. Unable to reach new hosts by casual contact, the germ had to find a warm refuge and a new means of transmission. It retreated to the mouth and, secondarily, the genitals. Spread mostly by common eating utensils and kissing, it caused a disease called bejel or nonvenereal syphilis. Now it attacked not only the skin and bones but the heart.

Venereal syphilis evolved from bejel perhaps 6,000 years ago, in the ancient cities of the Middle East. Now *T. pallidum* depended on coitus for transmission, and it might have to linger inside a person for months or even years before passing to another host. It survived by hiding in the heart and brain long after the initial skin lesion had disappeared, silent at first but wreaking slow damage. The treponeme may have mutated to greater virulence in the late fifteenth century, when the disease was first described in Europe, or perhaps it took a harsher toll as ships carried it worldwide to unfamiliar hosts. In the 1950s it seemed that penicillin might wipe it out, but changes in sex behavior and lifestyles gave it a new lease on life, along with the germs causing gonorrhea, chlamydia, genital herpes and AIDS.

The spirochetes of pinta, yaws, bejel and syphilis cannot be told apart by microscope or most lab tests. Experts long debated whether they are four strains of one germ* whose effects vary with climate and transmission, or whether they are subspecies of *T. pallidum* (the common current view). Regardless, it is clear that the lives of *T. pallidum, L. interrogans* and Bb all changed because people changed, and in turn they have affected people. Like the others, Bb sometimes survives for a long time in its hosts, causing chronic illness despite intervals of silence. All three can attack many organs but have a strong attraction to hosts' skin and nervous system. Like *T. pallidum,* Bb can affect the eyes, heart and joints. Before antibiotics, doctors said that to know syphilis was to know medicine, because it is an elusive chameleon that affects many organs. The same might also have been said then of Lyme disease. In fact, *T. pallidum* and Bb are so closely related that people with Lyme disease may score false positives in blood tests for syphilis.

By *Bergey's* last count, the genus *Borrelia* contains twenty-one species. They are less tightly and uniformly coiled than leptospires or treponemes, and most live in ticks that transmit them to mammals and birds. As they were discovered, they were assigned to species according to the

---

* A strain comprises all the descendants of one ancestral culture—say, the offspring of spirochetes grown in a lab culture from one sample of water or blood. If a number of strains share distinctive features, they constitute a species.

kinds of ticks they lived in, on the principle that a germ develops unique traits as it adapts to a particular host. Indeed some borrelia do inhabit only one species of tick and one larger host. One species infects poultry, another sickens cattle and horses, and another induces bovine abortion. These borrelia can infect humans but very rarely do so, because their host ticks almost never bite people.

The present list of twenty-one species will probably change. Now it is known that some borrelia, including Bb, can live in more than one tick. Furthermore, genome maps are revealing unsuspected similarities and differences among borrelia. For instance, three species, each carried by a different kind of tick, seem so similar genetically that they may be reclassified as variants of one species. And some varieties of Bb, such as those found in Europe, Asia and our northeastern states, are different enough to cause somewhat different symptoms, especially in chronic infections. As a result, some experts say that B. burgdorferi should be broken down into four or more distinct "genospecies." Time and greater knowledge will decide the issue.

However classified, borrelia cause three human diseases, Lyme disease and two types of relapsing fever (RF). RF flares acutely at first, causing a high temperature, jaundice and sometimes neurological symptoms; these fade after a week, only to return a week or two later. There may be three, four or even ten bouts of symptoms before the disease fades out. Each relapse occurs because the spirochete, under attack by its host, produces new surface proteins that baffle its immune system. It is one of the trickiest camouflage jobs in the bacterial world. Bb shares with its RF relatives some of this notable elusiveness.

Tick-borne RF is caused by three *Borrelia* species, all transmitted by soft ticks. It is quite rare in North America, usually the result of a tick bite in an abandoned barn or country cabin. The other type of RF, the louse-borne variety, is caused by *B. recurrentis,* the only species of *Borrelia* not spread by ticks. It makes its home in human head and body lice, and people are its only hosts. Louse-borne RF leaps up like wildfire when crowds of people are underfed, underwashed and overcrowded; between epidemics, it hides in some still undiscovered animal or tick reservoir. Like typhus, which thrives in the same conditions, it has caused much suffering and death during famines such as Ireland's Great Hunger.

So there are important differences among the spirochetes that cause human disease—the leptospire's hardiness, the treponeme's fragility, and the borrelia's need for ticks as carriers. But there are also important family likenesses. All three are clever at altering their surface proteins to confound hosts' immune systems. All can cause long chronic infections. All have some degree of affinity for brain tissue and can cause neurological symptoms. None produce toxins; many of the symptoms they cause result from hosts' immune reactions. People get sick trying to resist them. And these spirochetes have all had their lives changed by the effects of human technology and behavior.

NINE

Gaia, or Nearly Everyone's Cousin

A FINAL NOTE ABOUT BB'S PLACE in the Linnaean tree. The spirochetes' branch and borrelia's twig are far removed from ours, but that probably belies a hidden closeness. We not only belong to the same broad ecological community, we may be related at the cellular level. The Gaia hypothesis says that present-day spirochetes are direct descendants of other spiral forms that took up permanent lodging long ago in animal cells. If that is true, some of Bb's closest relatives are literally part of us.

Gaia grew from an idea put forth almost a century ago by a Russian biologist named Konstantin Merezhkovsky. He said that chloroplasts, the organelles that regulate energy production in plant cells, were once free-living bacteria. Perhaps they were something like the ancient bacteria that first performed photosynthesis. When they invaded plant cells, said Merezhkovsky, the hosts survived the assault but failed to kill the invaders. In time, invader and host reached a symbiotic peace, each performing vital functions for the other. The smaller partner took over energy production for both of them but lost most other abilities. The host cell's nucleus took care of their other needs. Now neither one could live without the other, and the one-time attacker had become an organelle.

This theory led to the idea that mitochondria, the energy-producing bodies in animal cells, are also relics of ancient invasions. Perhaps small, oxygen-burning bacteria entered larger cells, survived and ended up sharing the hosts' metabolic tasks. In the 1950s, an English theorist named James Lovelock created Gaia, the concept of life on earth as one huge community. All living things, he said, have interlocking needs and functions, and they act like a single superorganism. American biologist Lynn Margulis has elaborated the idea, stressing species' interdependence and cooperation. She claims that symbiosis is not, as was once thought, a curiosity of nature. It is common, in fact ubiquitous, and it helps to drive evolution.

Every type of animal cell, Margulis says, began with invasions; each organelle is evidence of an infection that became a partnership. A complex cell is really a community, consisting of a former host and its one-time invaders—in some kinds of human cells, as many as eighty of them. Some of the most important invaders, she believes, were spirochetes; they evolved into the hairlike cilia that give many cells mobility, and they may have assumed other functions as well.

At first such ideas lay at the fringes of science, but now a lot of evidence supports them. It has been learned that mitochondria have their own DNA, which divides on its own timetable, independent of the nucleus. Furthermore, symbiosis turns out to be common, occurring in many forms that seem to dramatize the Gaia theory. For instance, a protozoan called *Metopus* harbors two tenants, a bacterium and a methane-producing archaeobacterium; though a single creature, it is also a community of three different kingdoms of life: prokaryotes, eukaryotes and archaea. The protozoa causing malaria must have begun as such communities; they have organelles of both plant and animal origin. In recent years, researchers have not only observed such symbiosis, they have created and undone it in the laboratory. If the one-celled creature *Euglena* is treated with streptomycin, it is "cured" of its chloroplasts, and thus of its ability to perform photosynthesis. And if amoebae are infected with certain bacteria, they soon become so dependent on these occupiers that they cannot live without them.

Gaia began as an ecological theory about the mutual dependence of life-forms. Some dry-minded people reject it out of hand because it threatens to awaken their sense of awe. On the other hand, it has acquired a slightly ripe intellectual odor from some of its enthusiasts and popularizers, who confuse it with rapturous oneness with nature. I, too, have felt ecstatically close to certain living things, but I stop short of calling the biosphere one big pulsating organism, and pulsating with it. I refrain with affectionate skepticism;

while some of the "soft" versions of Gaia are devout and loving, they are more an inner state than an idea. Therefore they can produce reverence at the expense of understanding. I think nature offers enough mystery without one's cultivating more. In fact, I sometimes suspect that the fuzzier forms of Gaia are so popular because they give a sense of worshipful connectedness to people who are uneasy with God or formal religion.

I find the original, narrower version of Gaia warm and engaging enough, with its idea that we and such small, relatively simple creatures as Bb are cousins under the skin. It is also consistent with present-day genetics, which says that we share a core of common genes with all our evolutionary ancestors, from chimps through ticks and right on back to archaea. This means that our genes are more than a template for self-replication; they are a history of life on earth, passing down from species to species our common family traits.

TEN

Very Small Indeed

IT IS TIME, SURELY, to give Bb's physical portrait. One must first say that it is very, very small—so small that to most people the precise size is meaningless, a tedium of numbers to the right of the decimal point that defy the mind's eye. Most bacteria are measured in microns, or millionths of a meter (also called micrometers), a unit as unreal to common experience as the light-year. To put it another, equally unhelpful way, there are two and a half centimeters to an inch, and a micron is a ten-thousandth of a centimeter. *Borrelia burgdorferi,* the longest and most slender of all borrelia, averages 20–30 microns long and 0.2–0.3 microns wide, too narrow to be visible in the lower range of a light microscope.

Some writers try to make such tinyness real by translating it from metric to English measure; they give Bb's length as four-millionths of an inch, which seems no more real than 30 microns. Or they may try to make its dimensions mentally visible with arithmetic tricks, saying that 1,500 Bbs must be laid end to end or 100,000 side by side to make an inch. You might as well describe the distance to the moon by saying that it equals 200 million basketball pros laid end to end or twice that number of midgets. It would be just as true, and equally meaningless. I will not even try to fiddle with the next smaller measure after the micron, the millimicron (or nanometer), which is one-billionth of a meter.

The best way to get a sense of Bb's dimensions is to look first at small, familiar objects under a microscope and then, through increasingly strong lenses, at human body cells and microbes. One should start with everyday things—a grain of salt, a human hair, a cotton thread, a transparent slice of onion, a moth's antenna, a drop of pond water; one can go on to look at prepared slides of tissues from animal and human bodies and finally at bacterial cultures. These first views of the microscopic world are as awesome and humbling as one's first sight of the sky through a telescope.

I confess that there is some personal history tied up in my enthusiasm. I was thirteen when my basement chemistry and photography lab also became a microscopy workshop; I began to study and photograph butterflies, insect parts, pollen, protozoa, anything that would fit under the lens. Later came dissecting tools, a hand microtome for sectioning plant and animal tissue, and slides of stained human cells. I owed this to a generous uncle with a medical supply business who gave me a better microscope than a kid could hope for. I still have it, a massive old nine-pound, brassbound Bausch and Lomb honey with nine combinations of lenses, very clear magnifications up to 650× and slightly murkier ones to almost twice that. In the early fifties it was already a bit old-fashioned for professional labs, but it was good enough to make any student rejoice.

With that microscope I discovered the subvisible world as one should learn any art or science, by excited trial and error. There was mystery and exhilaration in the view through those lenses, and all my life—as a writer, a researcher, a psychoanalyst—my work has involved discovering things the eye cannot see. A Freudian could make much of that, and several already have, but I know my excitement is not just a personal quirk. I have seen the same fascination grip others when they realize firsthand how a microscope or telescope expands the visible world.

No one knows who first had that pleasure. The ancient Romans made lenses, and spectacles were invented in Europe around 1300. By 1600 lens grinding was much improved, and lenses were being aligned in tubes to reveal the heavens. Galileo, a great pioneer of the telescope, rearranged his lenses to look at small objects; he said they made flies look as big as lambs. At about this time, the first microscopes were being made in Italy and Holland. A half-century later, a draper in Delft named Anton van Leeuwenhoek began using them as Galileo had used the telescope, to reveal worlds no one had imagined.

Leeuwenhoek had no scientific or scholarly training; perhaps he started working with lenses to study the fabrics he traded. He became a very expert lens grinder, no small achievement when most glass contained many tiny air bubbles and was discolored by chemical impurities; some of the best lenses were ground from the pure quartz of a single grain of sand. What Leeuwenhoek could do with such tiny lenses is still amazing. The compound microscope, with large lenses mounted in a barrel and a mechanical stage for manipulating objects, was not perfected until the early nineteenth century. Leeuwenhoek's instrument was a small, hand-held affair made of brass and fitted with one lens the size of a pinhead. Next to a modern instrument it looks primitive, but with it he could get magnifications of several hundred, which was enough to reenvision the world.

In 1674 Leeuwenhoek used one of his little lenses to gaze at slime from a lake and to his amazement saw countless tiny animalcules spinning and darting about. For the next fifty years, he ventured into the invisible world with endless energy and inventiveness. He looked at pond scum and gutter water, fungi and plant fibers, his own blood, saliva, tooth scrapings, semen and feces. Everywhere he saw beautiful, intricate structures and boundless life. He was the first person to see protozoa, yeast cells, capillary circulation, red blood cells, striated muscle and spermatozoa. He was also the first to describe bacteria shaped like rods, spheres and spirals. Probably no one in history, not the first deep-sea divers nor the first astronauts, saw so many different things for the first time. Leeuwenhoek explored the subvisible world as Galileo explored the stars, Freud the mind and

Lewis and Clark the West. All of us who study life today walk in his footsteps. The best of us become, like him, amateurs in the first sense of the word, lovers of our subject. Even across three centuries, the enthusiasm of discovery that shines through in his letters is infectious. Leeuwenhoek, like John Aubrey, is one of the people I would most love to snatch back from the dark for one evening's conversation.

Leeuwenhoek wrote no books; he spent his long life grinding lenses, observing new marvels and sending his meticulous notes and drawings to the Royal Society in London. It was his English contemporary Robert Hooke who exposed the microscopic world to the public. Like Leeuwenhoek, he turned his lenses on everything around him—a razor's edge, the point of a needle, hairs, seeds, molds, gnats and fleas. Magnified one or two hundred times, the most trivial object became astonishing. A razor's edge was not even, it was as jagged as the Alps. Cork was not solid and amorphous, it consisted of rows of minute, air-filled chambers, which Hooke called cells because they reminded him of bare little rooms in a monastery. And there was his famous flea. In 1665 Hooke published *Micrographia*, one of the great science books of all time, with its magnificent drawings of microscopic life, including an oversize, fold-out picture of a flea. The flea was no longer puny and insignificant, it was like a miniature, many-legged rhino equipped with armor plate, sensory bristles and complex parts for grasping and feeding. To this day, Hooke's flea is an obligatory illustration in books on the microscope, as impressive as when it appeared and nearly as startling.

More than a century passed before it was understood that all life consists of cells, and a century more before microbes were related to disease. But thanks largely to Leeuwenhoek and Hooke, the microscope became a popular wonder and an indispensable research instrument. Samuel Pepys, among other curious laymen, hurried to buy *Micrographia;* he sat up with it until two in the morning and wrote in his diary that it was "the most ingenious book that ever I read in my life." In that era anatomists such as Eustachius, Fallopius, Malpighi and Swammerdam* began to use microscopes to explore the human body, and many of them left their names on what they discovered there.

You can still get a grasp of microbial dimensions by looking through a microscope at what lies around you. Start at low magnifications, with common objects that are small but visible to the naked eye, such as grains of salt; magnified fifty or a hundred times, they become great glistening cubes of crystal. Then, like Leeuwenhoek and Hooke, look at

---

* The microscope opened vistas, real and imaged, that were truly dizzying. With the discovery of sperm cells and ova, a long debate began about which of them contained the homunculus, the tiny, preformed infant thought to slumber within one of them. Logic said that each homunculus must contain still smaller homunculi, and so on ad infinitum. The great Dutch microscopist Jan Swammerdam had fits of near-ecstasy contemplating an infinity of ever-diminishing infants packed in each ovary or testicle, like so many nested Russian dolls.

the point of a pin or the edge of a razor blade, examine hairs and threads, and view the gorgeous frond of a moth's antenna. At 100–150 magnifications, motes suspended in pond water will be revealed as tiny multicelled animals, some of them wonderfully complex. At 200×, you can discern the cells of algae or of a thin slice of onion. And protozoa can be seen in pond water, the darting, slipper-like creatures and spinning whirligigs that delighted Leeuwenhoek.*

When you raise the magnification above 200×, you have left far behind even the smallest objects visible to a strong, youthful eye. Now you see some details of larger plant and animal cells. At 300–400× and above, you can study some of the internal differences among algae, protozoa and the larger cells of the human body; you may make out their nuclei and perhaps some organelles. You might conceivably glimpse Bb, but you would not recognize it; it is still just a speck too small and too faint to be identified. As you pass 500×, you have a chance of spotting Bb as a small but distinct spiral. However, because it is so slender, you may have to stain it with chemical dye or use dark-field microscopy, a

---

* There are exceptions to most rules of nature, including the one that protozoa and bacteria are invisible. The very biggest protozoa are visible specks. The behemoth of bacteria, *Epulopiscium fishelsoni,* was discovered in 1985 in the intestine of a surgeonfish in the Red Sea. It is more than a fiftieth of an inch long, almost the size of a small printed hyphen, by far the biggest bacterium ever seen.

trick of lighting that makes spirochetes glisten white against a black background.

At 600✕ and above, you have a better chance of identifying Bb when you meet it, and you begin to get an idea of its size in relation to other bacteria and to human cells. For instance, our disk-shaped red blood cells measure about 10 microns in diameter, a handy number for mental reference. Most bacteria are far smaller; the cocci, or spherical bacteria, commonly found in our bodies are only half a micron across; at 600✕ they are just minute dots. Bb, 20 to 30 microns long with its coils intact, is the length of two or three red blood cells side by side. However, its width, two or three tenths of a micron, make it seem frail by comparison; next to a red blood cell it is as thin as a line drawn with a sharp pencil.

To see Bb with ease and clarity, you must go to at least 1,000✕, and preferably to twice that. At 1,000 magnifications you enter the light microscope's high range, but most bacteria still look like opaque dots, dashes and squiggles, without internal structures. The microscope's oil-immersion lens may go as high as 1,300✕, but even then you cannot make out the smallest bacteria at all, and you certainly cannot see Bb's interior. To see its internal workings, you must use the electron microscope. This wonderful instrument, developed in the 1940s, employs an electron beam instead of light, and it can magnify tens of thousands of times. One recently developed type of scanning electron microscope magnifies up to 200,000 times. When using such an instrument, one measures in nanometers, or billionths of a meter, units so small that they blow away the sense of scale learned with a light microscope.

The electron microscope can reveal not only bacteria but viruses, chromosomes and even large single molecules. At 40,000× it makes Bb's insides clearly visible. At 60,000× it can show you in detail a cross section of one of Bb's flagella. Studying parts of one flagellum on a bacterium is like scrutinizing one hair on the head of someone in another town. This is not as startling as it once was, now that satellites can read license plates from their orbits in space. Still, it is awesome to see the electron microscope make visible the processes by which an individual cell lives. Without it, the portrait of Bb would remain the silhouette of a corkscrew.

ELEVEN

Not Just a Corkscrew

MOST BOOKS SAY *Borrelia burgdorferi* is a helix 20 to 30 microns long and leave it at that. They do not mention, let alone explain, that Bb's size and shape vary. Some strains have pointed ends; others are truncated or blunt. In a kindly environment Bb has a fair range of dimensions, spiral turns and flagella; in hard times, it is a virtual Lon Chaney of microbes—not quite a germ with a thousand faces, but one of vast plasticity. A microscopist who expects reality to match his textbooks could cross Bb's path ten times without recognizing it. There are, of course, reasons for Bb's shape-shifting. What Louis Sullivan said of architecture is true of microbes: Form follows function.

Even without the shape-shifting, Bb may be difficult or impossible to distinguish from some of its relatives by just looking. There is variety within each type of spirochete, and some overlap among types. Consider just the three genera that infect humans. Standard descriptions say that leptospires are 6–24 microns long, with 18 or more very tight, regularly spaced coils. Treponemes are the same length but have only 6–14 coils, which are usually rather tight and regular, but not always. Bb averages 20–30 microns, and like most borrelia has 3–10 loose, irregular coils.

These pictures do hold as generalities, but Bb can be as short as 10 microns or as long as 40; its length depends on individual variation and on whether the germ stretches out like a thread or contracts like a bedspring. Other spirochetes also vary, so sometimes it does not help to know that Bb is, on average, the longest of borrelia, or that it is less tightly coiled than a treponeme. Whether you can identify a given germ as Bb or some other spirochete may depend less on visual detail than on which species of tick it came from or whether it ferments glucose into lactic acid.

Furthermore, Bb can change more than its length and the number and tightness of its coils. Like many spirochetes, it can switch its shape from a spiral to a filament, cyst, granule, hooked rod or elbow. These variants are called L forms, a reference not to their shape but to the Lister Institute in London, where they were first studied. They are also called cell-wall-deficient bacteria; they take these nonspiral shapes when they have lost much of their cell wall. In harsh environments, many microbes do what large creatures do in famine or foul weather: they hunker down, reduce their activity, and try to get by on stored energy. This can be seen in lab cultures of Bb, when the germ first thrives but then falls victim to its own success. Overcrowding sets in, nutrients dwindle, and toxic wastes accumulate. Many of the microbes die, but others survive at smaller sizes, expending little of their energy on maintaining a wall. In such circumstances, Bb may turn into a minute granule; put in fresh growth medium, it turns back into a corkscrew.

Many L forms keep their normal abilities. Even when not overcrowded, Bb in a lab culture sometimes becomes a filament or cyst, and in these shapes it can still reproduce. The small, granular form of Bb can also infect hosts and cause disease; this may help explain why Lyme disease symptoms may persist even though Bb's spiral form has apparently vanished. L forms may also help explain why Lyme disease sometimes resists antibiotics, many of which work by attacking microbes' cell walls. The deficient walls of L forms may deprive these drugs of a target. If all this is true, the L form is a truly remarkable survival tactic.

L forms are a recondite aspect of microbiology, but they are strong evidence of Bb's versatility at infecting hosts and surviving inside them. Further details belong to the specialist, but anyone who comes across mention of Bb's L form should know that the term refers not to its shape but to its power of metamorphosis. I will continue to describe Bb as a minuscule corkscrew, but with the understanding that this is only its most common incarnation.

TWELVE

A Possibly Poignant Anatomy

MANY BOOKSHELVES HOLD a copy of *Gray's Anatomy* that has never been opened. Perhaps it sits unused next to other treasures, such as *Shepherd's Historical Atlas, Brewer's Dictionary of Phrase and Fable* and the small-type *Oxford English Dictionary.* Even unused, such books give a certain satisfaction. They were bought with good intentions, however vague, and the intentions and sense of respect remain. To keep them in sight, unvisited but not forgotten, is the homage of an uncalloused mind. Even though the owner has never entered the tangled realms of neuroanatomy, etymology and historical geography, he knows that should he ever want to, he has excellent guides to follow.

A great deal has been written about Bb, but a *Gray's Anatomy,* a single definitive work, is probably still decades away. Even if I could suggest such a book, most people would probably leave it on the library shelf. They would no more wrestle with all of Bb's outer-coat proteins than with all the ligaments of the human metatarsal bones. Fortunately, our purpose demands only a basic portrait of Bb. What follows is a picture of the microbe as seen from the middle distance—more than the light microscope reveals, but not a detailed molecular view. That is, it consists mostly of features that can be seen through an electron microscope at about 50,000 magnifications. It begins with Bb's insides and proceeds outward.

Bb is basically a long, slender cylinder of cytoplasm, the living stuff of every cell's interior. The cytoplasm is surrounded by a membrane, a wrapping of flagella, an outer wall and a layer of slime. Embedded in the cytoplasm are organelles, fewer than exist in plant and animal cells but sufficient for Bb; small bodies called ribosomes and mesosomes carry out many of its metabolic tasks. As in all bacteria, there is no walled-off nucleus; most of Bb's DNA lies in a vague nuclear region, in a single chromosome. This is slight genetic equipment compared to that of an animal cell; the malaria protozoan has 14 chromosomes, a human cell 46. Even as bacteria go, Bb has a dwarfish genome; 853 genes lie on the chromosome, and another 430 are contained in a dozen or more plasmids, bundles of DNA scattered about the cytoplasm. These genes, fewer than 1,300 in all, carry the entire operating code for Bb's life, directing its digestion, metabolism, reproduction, weapons of attack and defense, and responses to a changing environment.

Borrelia are among the very few bacteria that have a

straight rather than circular chromosome; some of their plasmids are also linear rather than coiled. Why they have this shape is unknown, and probably important. We do know that the plasmids contain duplicates of some genes on the chromosome; apparently this lets Bb turn them on and off in a greater variety of sequences, altering outer-coat proteins and confounding hosts' defenses. Some of the genes on plasmids control Bb's ability to attach itself to a mammal's tissues, and others provoke symptoms in hosts. We know this because Bb cultured in a lab tends to lose certain plasmids; with the disappearance of some it loses the ability to infect lab animals, and with others the power to make them sick.

Bb's cytoplasm is contained by a double wall built of proteins, complex sugars and lipids. To Bb this wall is the equivalent of a skin and senses; it registers changes in the outside world, sends this information to the genes, and carries out their instructions to respond. Some outer-surface proteins allow the selective passage in and out of nutrients and wastes, and others interact with a host's tissues and immune defenses. The wall can do all this because it is a multilayered system of molecular sieves and docking ports. It consists partly of peptidoglycan (PG), a substance built of proteins and sugars that is unique to bacteria. PG helps Bb maintain its shape and, incidentally, sets off immune reactions in hosts that cause fever and malaise. Because PG is vital to microbial life, it is one of the targets researchers aim at in developing antibacterial drugs. Penicillin and some other antibiotics are effective because they inhibit germs' ability to create PG and thus to maintain outer-wall integrity.

Most textbooks, having first described a germ's size and shape, then say whether it is gram-positive or gram-negative. Many chemical dyes are used to reveal cells' structure under the microscope; which dyes a cell absorbs and which it rejects can tell a lot about its chemical makeup. In 1884 Danish bacteriologist Hans Christian Gram developed a staining technique that is still used to classify bacteria; whether germs take it up depends on their walls' thickness and molecular structure. Gram-positive organisms, which hold the blue-violet stain, have thick walls with many layers of PG and some distinctive chemical components. Gram-negative germs, which reject the stain, have thinner walls with less PG and different outer-coat ingredients. Because of their different walls and surface proteins, gram-positive and gram-negative bacteria have different effects on hosts, and they are susceptible to different antibiotics. Bb is usually labeled gram-negative, but its response to Gram stain may not be perfectly consistent, so it has sometimes been called gram-variable. This may reflect the shiftiness of Bb's outer-surface proteins, which play hide-and-seek so well with hosts' defenses and with antibacterial drugs.

Bb's outer wall is surrounded by a layer of slime, a feature it shares with some other bacteria, such as a number of spirochetes and the TB bacillus. This slime consists mostly of complex sugars, and apparently it helps protect the germ from hosts' immune systems, especially from being engulfed by roving white blood cells—or, if it is engulfed, from being digested. The slime layer may also help spirochetes cling to such slippery surfaces as tooth enamel and mucous membranes, and to resist being flushed from their anchorages by the host's body fluids.

One of Bb's most remarkable features is the whiplike flagella that grow from both ends. There are usually seven to eleven flagella, but sometimes as few as four or more than a dozen. Bb's flagella, unlike those of most microbes, do not project into the surrounding fluid to row it about. They arise in the cytoplasm, pass through the inner wall, and wrap themselves lengthwise around Bb's body in the narrow space between its inner and outer walls. This makes Bb rather like a curvy maypole with its streamers wrapped around it. Each flagellum consists of protein threads intertwined like strands of rope, and it is made to rotate by a sort of biological motor at its base. This turning of the flagella is what makes Bb move snakewise through liquids or turn like a tiny screw through the spaces between hosts' cells. Flagella also contain a protein, flagellin, that has strong and unpleasant effects on hosts.

The mapping of Bb's genome in 1997 has opened the way for a more detailed picture of its anatomy and physiology. Bb is at home deep in hosts' tissues; it thrives best when it gets very little oxygen. It burns glucose, a carbohydrate, as its major energy source, and it produces lactic acid as a byproduct. However, Bb seems to entirely lack genes to direct certain basic metabolic processes, such as creating amino acids (the building blocks of proteins) and nucleotides (the building blocks of DNA). That is why one cannot grow Bb in a laboratory culture medium without feeding it dozens of vitamin and protein supplements; normally Bb draws these vital materials from hosts rather than manufacturing them. This is also true of T. pallidum, the syphilis spirochete, whose genome was mapped within months of Bb's. A current theory holds that both spirochetes evolved from a more complex common ancestor, and that they lost the genes for creating certain substances when they learned to hijack them from hosts. *

While Bb's outer proteins and metabolism are fascinating, further details about them belong in the province of full-time biologists. Besides, with molecular biology dashing forward so quickly, much that one writes now about Bb's physiology will soon be outdated. A great deal of it must be written on water, not carved in stone. That is the fate

---

* Evolution was once equated with progress and increasing complexity; life seemed to march up, up, up, from primordial jellies to the Victorian upper classes. Now it is thought that some species succeed in life by going backward. Viruses were once considered a half step between inanimate matter and cellular life; certain minute microbes (rickettsia and mycoplasmas) were thought to represent a primitive stage between viruses and bacteria. Now these tiny microbes are seen as stripped-down versions of more complex ancestors; they shed all the internal machinery they could and borrowed from hosts. Bb and T. pallidum may have lost some genes this way.

of most science, to become obsolete. Few books are more touching than yesterday's authoritative text, with all its quaint convictions. Sometimes looking at an earlier era's ideas about genetics or reproductive behavior is like reading yesterday's melodrama; you wonder that anyone bought it.

This lasting impermanence does not discourage scientists. They accept with surprising equanimity that a new idea or a new instrument may suddenly turn their theories into antiques. If not, increments of knowledge over time will probably do so. Yet they work on, actually hoping to be superseded. That is what makes science such a fragile, poignant enterprise.

THIRTEEN

Instead of Sex

ONE OF THE BEST of Robert Benchley's droll little essays was "The Sex Life of the Amoeba." At the height of his popularity it was made into a movie short, with the dapper, quizzical Benchley himself speaking the piece. A microscope stood on his desk, but only Benchley got to look through it; the audience never saw what he claimed was happening. Had they been able to peek, there would have been no humor; the laughs came from Benchley's show of embarrassed titillation. That, I fear, is the best I could do with Bb's sex life, for it has none. But Bb, like all living things, must reproduce, and it does so in several ways; what it lacks in copulatory drama, it makes up in variety. And there is more to asexual reproduction than scientists used to think. It is not just a primitive precursor of sex, but a different and very successful reproductive strategy.

All plants and animals reproduce sexually; two individuals mix their genes rather than just cloning themselves, as bacteria do. Why they do so is a hot topic in evolutionary theory; it was long said that higher species endure all the requisite conflict and heavy breathing because sex promotes genetic diversity and thus adaptability. Or perhaps, a more recent theory says, greater diversity was a response to parasitism; sex may have first evolved as a way to fight infection.

Whatever their reason, bacteria have ignored it. Taking an apparently simpler path, they reproduce asexually; the single chromosome replicates, and the cell divides into two virtually identical halves. It used to be assumed that genetic variation came rarely, through spontaneous mutations alone. Now we know that bacteria are genetically much less static than plants and animals. Just as microbiology discovered a tumult of proteins on microbes' outer walls, it revealed a constant buzz of genetic traffic within and between microbes.

That traffic takes many routes. Phages, the viruses that infect bacteria, often inject their genes into a host's DNA, sometimes their own and sometimes genes kidnaped from earlier hosts. Also, bacteria can absorb floating fragments of DNA from their surroundings, even the genes of other bacterial species; that is how transposons, or jumping genes, get from one microbe to another. Sometimes entire plasmids travel between bacteria, creating new traits in the recipient. And finally there is conjugation: one bacterium joins itself to another by extending a rigid, hairlike tube called a pilus, and through it donates DNA from its chromosome or plasmids. Conjugation looks rather like sex, but the process is one-way DNA transfer, not true gene mixing. It seems to be a precursor of sex.

Because virtually every organism bigger than a virus is parasitized, genetic traffic is ubiquitous, and its effects are powerful. One jumping gene can give a bacterium immunity to antibiotics or alter its virulence; the inclusion of a wandering plasmid may create, in effect, a new species. For instance, the emergence of a new strain of *E. coli* causing dangerously severe food poisoning results from this species picking up a toxin-producing plasmid from another type of bacteria. So despite their lack of sex, bacteria have more genetic fluidity and variety than sexual species. That is what gives them their amazing versatility and evolutionary success. It remains an intriguing, unanswered question how much genetic traffic goes on between bacteria and higher species.

Bacteria reproduce at quite different speeds. *E. coli* splits in half every twenty minutes, *T. pallidum* every thirty hours; Hansen's bacillus, which causes leprosy, divides every two weeks. Seen in that continuum, Bb seems neither fast nor extremely slow—it divides every eight to twelve hours. Reproductive speed is no small matter, because a germ's entry into a host is always the start of a race that pits the germ's ability to multiply, colonize and spread against the host's ability to mobilize antibodies and other defenses. The loser may perish, so a less than speedy multiplier such as Bb must have a deceptive outer coat and other ways of buying time.

Bb often multiplies as most bacteria do, by binary fission. However, like a number of other spirochetes, it sometimes does so in quite different ways. Sometimes it develops large cystlike bodies in which small spirochetes develop, which are then released. Or buds may form along Bb's length; each one eventually becomes a cyst from which new spirochetes emerge. Sometimes Bb sprouts branchlike filaments, each of which becomes an independent spirochete. Even Bb's odd little L forms can reproduce. How all these reproductive variations affect hosts or are affected by them is still to be discovered.

One unappreciated aspect of Bb's reproductive life is that often it multiplies while doing other, quite difficult tasks. Where and when to reproduce are matters of self-preservation for any creature. Whether reproduction involves reshaping a cell wall or risking the oblivion of orgasm, it means vulnerability—to predators, competitors and environmental stresses. Humans, like many other species, tend to avoid such risks; it is not just modesty that makes lovers retire to secluded places. In virtually all societies, most people usually avoid coitus in exposed places; rather than using the village green or city hall steps, they retreat to some sheltering bower, or at least a motel. Bb, on the other hand, manages to reproduce where and when it must, while fending off a host's cellular and chemical attacks. This equivalent of dueling while making love seems to have no human equivalent. Surely it defies Zoltán Kodály's words, spoken to commissars who challenged his music's lack of social uplift: "You cannot make love and read the newspaper at the same time."

FOURTEEN

A Fantastic Voyage

IN 1966 *Fantastic Voyage* appeared, a science-fiction film with a premise worthy of Jules Verne. A political defector from a foreign power, having barely survived an assassination attempt, lay comatose in a secret laboratory in Washington with a blood clot pressing on his brain. To save him, medical and military leaders joined their expertise and secret arsenals: they put five people in a submarine, reduced sub and crew to "about the size of a microbe," and injected them into the patient's carotid artery. This launched them on a wondrous voyage through inner space, to find and destroy the clot. They fought their way through the patient's heart, lungs, lymph glands and inner ear to reach his brain. Along the way they dodged antibodies and white blood cells, became entangled in neurons, and were blown about by respiratory hurricanes. Finally, having used a laser to dissolve the clot, they abandoned their sub, trekked along the optic nerve and swam to freedom in a teardrop. Despite some cartoonish lapses—body maps like plumbing charts and Raquel Welch as backup neurosurgeon—it is cracking good adventure and anatomically crudely accurate.

It is such a voyage I envision for *Borrelia burgdorferi* as it travels through its several hosts (for convenience I speak of it in the singular, though of course it exists only as one of a crowd). Along the way it faces perils much like those in *Fantastic Voyage*; in fact, it must survive more obstacles than Hollywood imagined, and defenses far more ingenious. Each host Bb inhabits, from tick to human, is bigger, stronger and vastly more complicated than itself, yet none is more resourceful nor ultimately more successful.

Bb's travels start in a tick, which does not make my task easier. If germs are a hard sell to some people, ticks are harder. They arouse as much hate and revulsion as leeches, and sometimes the sort of anxiety stirred by rabid bats. Aristotle called them disgusting, and much modern opinion is no higher; if you sing the praises of ticks' ingenuity, people may stick their fingers in their ears. The fact remains that the tick, like Bb, is a compact marvel of adaptation. And of all the many varieties, none is more finely tuned to its hosts' bodies and habits than *Ixodes scapularis,* the deer tick, or black-legged tick, which is Bb's primary home.*

---

* This tick is the subject of one of those Linnaean scuffles I mentioned; it can confuse nonspecialists. In 1979 Harvard entomologist Andrew Spielman claimed that the tick carrying Bb in the Northeast is not *I. scapularis* but a separate species he dubbed *I. dammini.* For a while this

Ticks, like spirochetes, have a lineage that probably goes back several hundred million years. This versatile parasite is not an insect but an arachnid, a class that also includes spiders, mites and scorpions. Ticks and mites form their own subclass, called acarids, and they are quite different from spiders. Spiders are free-living carnivores, some of nature's fiercest little predators. Ticks, by contrast, are true parasites; instead of hunting for food, they cling to hosts' skin and hijack their body fluids. There are two types of ticks, soft and hard, with different ecological niches, life cycles and feeding habits. Soft ticks, such as the one that transmits the borrelia of tick-borne relapsing fever, live in caves and abandoned cabins. Since potential hosts may stumble in only at very long intervals, these ticks must be able to live for years without eating. When they do feed, they fill up with greedy speed, in an hour or less. Hard ticks, such as *I. scapularis,* live in meadows and scrub vegetation and on forest floors; to survive there, exposed to weather and predators, they need the tough protective shields that cover their backs. They take at least a few days to feed, and these long meals cannot be completed without some remarkable skills.

As acarologists have discovered, the deer tick has a two-year life cycle with three stages; at each stage the tick must

---

view was widely accepted; then other researchers argued that the two species are one, and the original name was restored. That is why many writings from 1979 to 1993 call Bb's carrier *I. dammini,* but earlier and later ones speak of *I. scapularis.* The last word on this may not be in.

drink a blood meal or die. It hatches out in midsummer as a six-legged larva about the size of a printed period and immediately starts seeking a host. This first host is usually a small mammal, such as a mouse or squirrel, or perhaps a small ground-foraging bird, such as a robin or blue jay. In the northeastern United States, the larval tick's most common host is the white-footed mouse; rarely does it feed on anything as big as a deer or human. After attaching itself to a mouse, the larva drinks blood for two or three days, tripling in size; then it drops off and lies quiescent through the winter. In spring it molts into a somewhat larger, eight-legged form called a nymph.

Despite its name, the nymph is an unlovely little creature with a big blood thirst; it may feed on a mouse or something larger, such as a rabbit, fox, dog or human—it is mostly nymphs that infect people with Bb. The nymph drinks mightily for several days, increasing its weight up to a hundred times; then it falls to the ground and lies dormant until the next summer, when it molts into an eight-legged adult. The adult attaches itself to a white-tailed deer or, for lack of one, some other big mammal. This final host gives the tick its name, but the deer is its home for only a week or two, as a platform for feeding and breeding. Mouse tick might be a better name, but deer tick probably stuck because only as an adult, feeding on deer, is *I. scapularis* big enough for most people to notice it. By then the tick is singing its erotic swan song.

Unlike larvae and nymphs, adult ticks have visible sex differences, and their lives center on reproducing. How they do so is worth telling, because sex is varied and fascinating even in ticks, and because almost everyone who writes about ticks and Lyme disease leaves it out. Actually, courtship and copulation are less flamboyant in deer ticks than in some of their relatives, but any tick's sex life is odd to human eyes. One thing, however, is familiar: ticks, like so many creatures, link the joy of sex and the joy of eating. Their mating is synchronized with their last meal. Previous feedings fueled a tick's growth and moltings; the final feast of adulthood energizes the male to produce sperm, the female to produce thousands of eggs, and both to copulate—which some of them do heroically.

In some common ticks, such as the dog tick, reproduction starts when an adult female begins a long, slow meal on her host. She emits a potent pheromone, or airborne hormone, that a male can pick up the tick equivalent of a dozen football fields away. He finds it irresistible and follows his nose, so to speak, across the host's fur until he reaches its source. His arrival stimulates the female to emit a whole series of pheromones, each one precisely timed to stimulate the next step in mating.

Those steps are not the ones people are familiar with; a tick's external genitals are little more than openings, called genital pores. The female does have a vagina of sorts, but the male lacks anything like a penis. He mounts the female and probes her genital pore with his mouth parts, in some species for hours. Finally he extrudes sperm from his genital pore in a little pear-shaped packet; he grasps it with his mouth parts and places it on the female's genital pore. Eventually the sperm enter her vagina.

The male's bliss, however strange to humans, must be considerable; given an opportunity, he may do it again and yet again, alternately feeding and mating, and doubling his weight in the process. In some tick species, he may repeat the act a couple of dozen times. At last even he gets tired; he falls off the deer and dies, without so much as a brief aria. The female tick may also mate heroically, but she pays more attention to food. In some species her copulation is timed to a binge of intense feeding that lasts twelve to thirty-six hours; acarologists, with straight faces, call it "the big sip," a final orgy of *gourmandise* that increases her weight a hundredfold. After eating and mating, she drops to ground, lies there till the following spring, and then lays her eggs and dies.

In deer ticks, unlike many other species, the male is not led to his mate by a pheromone; apparently he just wanders at random until he finds one. This seems inefficient for a slow mover like the tick, but it works well enough. Once he finds her, he is guided by the usual sequence of sex hormones. The male deer tick inserts his mouth parts into the female's genital pore for a long while, removing them only after his sperm have fully entered her vagina. Some males copulate just once, but others do so two or three times. Females, too, may mate more than once; the last mating will be the one that succeeds in fertilizing her eggs.

In the Northeast, deer ticks mate in the fall, and the female lays her eggs the following spring; in August a new generation of larvae emerge, and the cycle begins again. Depending on the weather and how quickly ticks find hosts at each stage of development, their life cycle may be timed a bit differently and take up to three years.

Once hatched, a tick spends most of its life finding hosts, a job it does like a virtuoso. Finding hosts is no mean task for a creature that has no wings, cannot jump, and walks like a nonagenarian. Rare is the tick that ever travels more than a yard or two from where it hatched. Its questing—acarologists' gallant term for seeking hosts—consists mostly of waiting. In exquisitely slow motion, the tick trundles to where hosts are most likely to brush against vegetation and pick up passengers, which is anywhere from several inches to several feet above the ground. Genetically programmed to move against gravity, the tick at each stage of its life trudges up stalks of grass or stems of shrubbery. Then it waits to seize a passing host with a pair of legs specially adapted for grasping.

At first glance, the tick seems almost too simple a creature to survive, let alone find and feed on hosts. Though endangered by birds and other predators, it lacks teeth, claws, mobility and camouflage. It even lacks real blood; a fluid called hemolymph carries nutrients and oxygen through its body. During its long quests, it is constantly threatened with death by dehydration; it must periodically labor its way down to the ground to find water and clamber back up to resume waiting. Unlike insects, the tick has no brain to guide it, just a large central ganglion located in its body. It does not even have a true head, just a sort of enlarged terminus atop the thorax, with eyes and mouth parts.

Despite these apparent deficiencies, the tick is wonderfully equipped to sense and seize hosts. It is covered with tiny hairs and sensory pores that respond to heat, light, humidity, touch, airborne vibrations and odors; they can also detect certain chemicals, including the carbon dioxide mammals exhale. When these multiple sensing systems detect a host's approach, the tick's grasping legs and ingenious body chemistry become poised for action. It grabs hold of the host's skin and slices out a small, neat wound. Its mouth parts are shaped not to tap small blood vessels but to create a little pocket into which blood will seep. The result is a constantly replenished little pool from which the tick can sip.

In any mammal, from mouse to man, such intrusions set off countermeasures that should suffice to baffle a tick. Damaged cells at the bite wound release histamines and other substances that cause inflammation and pain; these tell the host it is being attacked, moving it to try to dislodge the offender. Coagulants and fibrin form a clot to stop the bleeding. Cascades of chemical messengers are set off by the wound and by the tick's saliva; they summon white blood cells to attack and signal T cells to trigger the production of antibodies.

The tick is not overwhelmed, for it has its own chemical

arsenal, with a reply to each host defense. Its saliva contains a material that hardens in minutes to a rubbery plug, cementing the tick in place and frustrating efforts to remove it. The saliva also contains a local anesthetic and several anti-inflammatories, which inhibit swelling and itching; together these substances may keep a host from noticing that the tick is there. An anticoagulant keeps the host's fibrin and platelets from forming a clot, so blood keeps flowing into the wound. Prostaglandin E2, a hormone-like substance, is injected into the wound to inhibit production of antibodies against the tick and its saliva.

Even this arsenal could not protect the tick if it were used just once; since the tick must feed for days, its protection must last that long. Therefore saliva is injected into the wound in repeated pulses, as part of a fluid-control system that would please a Dutch hydrologist. The tick has to store enough nutrients for months, but its blood meal is mostly water; if it did not dump much of the fluid, it would burst. It solves several problems at once by concentrating the nutrients of a meal in its gut and sending excess water to its salivary glands, which flush it out. By alternating periods of feeding and flushing, the tick keeps bathing the attachment site with saliva and all its salivary weapons. Thus it can feed for days, swell but not burst, and drop off the host without even having been noticed.

Through long evolution, ticks' behavior and body chemistry have been finely tuned to those of their hosts. Few ticks are indiscriminate feeders; one species lives only on rabbits,

others only on birds, dogs or sheep. *I. scapularis* and others
with multiple hosts have developed internal clocks that syn-
chronize their hatching, questing, moltings and mating with
all their hosts' habits and life cycles. The so-called deer tick
is really most closely adapted to the white-footed mouse,
but it is very versatile; it can digest the blood of a chipmunk
or human and resist the immune system of a wren or raccoon. This demands a bigger range of adaptations than
most parasites must make, but it confers a major benefit: if
weather, famine or epidemics should deprive the tick of its
preferred host, it has alternatives other than death by starva-
tion.

Those many adaptations may serve the tick well, but they
compel Bb to travel back and forth between ticks and a variety of mammals and birds, meeting a daunting series of
obstacles. It replies to each one with ingenuity. In fact, the
intersecting of all these species' life cycles and defenses bring
to mind an improbable Rube Goldberg invention; it seems
almost too tricky to succeed. After all, the more complex
any mechanism is, the more can go wrong with it. Yet the
entangled relationships of Bb, the deer tick, the white-footed
mouse and the tick's other hosts do not break down in accident or failure. For all its intricacy, this is an ancient and
robust system, as Bb's fantastic voyages show.

FIFTEEN

Equally Fantastic

THE DEER TICK MAY BE Bb's most ancient and familiar home, but getting inside it is no cakewalk. Any germ's passage from one species to another is a leap into another country; biologists call it crossing a species boundary. Like the crossing of national and geographical boundaries, such a trip can be dangerous, even fatal. Depending on how and where a germ enters a host, it may be killed by any number of shocks, from the sudden change in temperature to attacks by salivary enzymes, digestive juices, white blood cells and antibodies. To have all the advantages of life inside a host, the microbe must marshal defenses against all these threats. Bb makes many journeys across species boundaries during its infective cycle, which includes all three stages of the tick's development and three other hosts. Despite its delicate looks and small genome, it has thrived while doing so.

Bb crosses the first boundary in early summer, when a newly hatched tick larva, free of borrelia, seizes a host and takes a bite. If the tick lives in the northeastern United States—say, a spacious backyard in Lyme, Connecticut—that host will usually be a white-footed mouse. The mouse became infected with Bb earlier that season, when bitten by a borrelia-laden nymph that had hatched the previous year. Since then the mouse has been carrying Bb in its blood and skin; as long as it lives it will be able to transmit the spirochetes to newborn ticks. Thus the mouse passes Bb from one generation of ticks to the next, receiving it from nymphs and giving it to larvae. In the language of epidemiology, the tick is Bb's vector, or carrier, and the mouse is its reservoir.

Inside a mouse, Bb is warm, well fed and at home with the rodent's immune defenses. But when a tick bites the mouse, everything changes. Bb is swept along in the blood flowing into the tick's mouth and down to its gut. It is as if a person were cast in one moment from a hot tub to a meat locker. Some spirochetes are too delicate to tolerate such shocks; the syphilis germ is so closely attuned to normal human body temperature that in pre-antibiotic days, some doctors infected syphilis patients with malaria, hoping the high fever would cure them. But Bb is hardy and versatile; it adapts swiftly to the cooler temperature inside the tick, and to taking its nourishment from hemolymph instead of rich mammalian blood. It also copes successfully with the tick's immune system, which is simpler than a mouse's but has its own unique weapons.

Bb can make such adjustments because its outer wall detects environmental changes and signals them to its interior; genes in the chromosome and plasmids respond by turning off certain proteins on Bb's exterior and activating others. These outer-surface proteins, or Osps, are vital to Bb's survival. In fact, much research into any germ today is about the molecular bustle on its surface, where small variations can have huge consequences. The difference between deadly *Yersinia pestis,* which causes bubonic plague, and some merely inconvenient relatives is probably just a few Osps. In *Borrelia,* too, outer-wall proteins determine whether a germ can survive entry into a given host, attach to the right tissue inside it, create a colony there, survive immune-system assaults and provoke symptoms.

Some of Bb's surface proteins are found widely in nature, but others exist in Bb alone. The first three unique ones were named, in very low flights of fancy, OspA, OspB and OspC, and they became targets in the development of Lyme disease vaccines. It was soon evident, however, that the production of each of these proteins is turned on and off at various points in Bb's life, so they are moving targets. For instance, when Bb travels from a tick to a mammal, it stops producing OspA and starts making OspC; it is OspC that alerts the human body to Bb's presence and provokes its first immune response. Also, various strains of Bb around the world have different amounts of OspA, OspB and especially OspC, which probably helps explain why they cause slightly different symptoms. Furthermore, OspB and OspC seem to go through changes in the course of chronic Lyme disease, confusing host defenses. Like the spirochetes of syphilis and relapsing fever, Bb is adept at Osp deception, a talent that made devising Lyme disease vaccines quite difficult. Vaccine and drug research has so far focused chiefly on OspA, but someday it may aim successfully at other surface proteins.

The workings of Osps are intricate and wondrous. One surface protein may have many functions, and sometimes several cooperate on a single task. For instance, several Osps (especially A and B) help Bb anchor itself to a host's tissue, a feat not to be taken for granted. To any microbe, the host's organs are as different as mountains, meadows and glaciers; neurons, intestinal lining and other cells have distinctive shapes and surfaces; germs attach to them with help from surface proteins called adhesins, which match up with host cells' molecular docking ports. After attaching to cells, some germs penetrate them and live in their interior; they do so with help from proteins called invasins, which have an affinity for host proteins called integrins. The study of microbial adhesion and infection has revealed an alphabet of Osps and given rise to an awkward poetry of molecular combat, replete with invasins, defensins, integrins, intimins and penetrins.

Yet another protein that helps Bb with attachment is flagellin; as its name suggests, it is a major ingredient of the flagella, and it causes some of Lyme disease's early symptoms. Bb's surface also has heat shock proteins (HSPs), which are activated by temperature shifts and other stressors, such as toxins and invading viruses. These proteins also help to build, fold and transport other molecules, a talent that earns them the description "molecular chaperones." Some varieties of HSPs probably exist in most cells in nature, but they are especially important to Bb, which in crossing so many species boundaries must endure frequent, varied shocks. HSPs, like flagellin, trigger defensive reactions in hosts that help to provoke the flu-like malaise of early Lyme disease.

So thanks to its surface proteins, Bb adjusts quickly to the abrupt leap from mouse blood to a larval tick's gut. With help from adhesins, it clings to the gut lining and resists being digested along with the tick's meal, as it would be if swallowed by a person or even a mosquito. It is death by digestion that keeps Bb from being transmitted by mosquitoes, a possibility people feared when Lyme disease was discovered.

Bb remains safe in the tick's gut because the tick does not digest food there, it just concentrates its meal and shunts away excess fluid. The concentrate is absorbed into the interior of the gut wall, which is where the tick digests it. Unmolested by digestive enzymes, Bb colonizes the gut lining, often from a very modest population base. Many germs have to invade hosts by the tens or hundreds of thousands to establish themselves, but a mere 4,000 Bb can colonize an adult tick, and several hundred may be enough to infect a larva or nymph. That is impressive infective power—to use a human analogy, it is as if the *Mayflower*'s passengers filled North America with descendants in a trice.

Having formed its initial colony, Bb penetrates the gut wall and multiplies in spaces between cells. There it stays, more or less active, until the tick feeds again, at the next stage of its life. That next blood meal will reactivate the spirochetes, allowing them to enter the tick's hemolymph. The hemolymph will carry them throughout the tick's body, and they will form colonies wherever they can. However, they have a special affinity for the central ganglion, ovaries and salivary glands. The salivary-gland colonies are essential for Bb's transmission, because the tick does not spread Bb by regurgitating it from its gut while feeding, as many biting insects do. Rather it injects Bb in pulses of saliva. It is a very efficient kind of transmission, the same one mosquitoes use to spread malaria, yellow fever and encephalitis.

But an infected larval tick cannot transmit Bb; it must let go of the mouse that infected it, fall to the ground, and spend the winter there. The next summer, it will molt into a nymph, quest and find its second mammalian host; perhaps this time it will again be a mouse or chipmunk, or maybe a raccoon, dog or human. Let us say it is a raccoon. The infected nymph bites it and begins to feed. This activates the Bb in its gut, and it injects the germ into the host's wound; again the spirochete crosses a species boundary. Now the chemicals in the tick's saliva help Bb; they neutralize the raccoon's immune defenses at the wound site, locally suppressing white blood cells, T cells and antibodies. Under their protective shield, Bb uses its heat shock proteins to adjust to the warmth and chemistry of the raccoon's body, and it employs other surface proteins to establish a colony there. Bb will perpetuate itself indefinitely inside the raccoon, which will be able to infect future generations of ticks and keep the infectious cycle going.

After several days of feeding, the nymph drops off the raccoon, passes the winter and transforms into an adult. The adult feeds on a deer or some other large mammal; it will mate on the deer and produce a new generation of larvae. These larvae pick up Bb from infected mice or other reservoir species, and the infectious cycle begins again. Once more Bb will survive several tick moltings and up to a half-dozen journeys between ticks and mammals, and perhaps a period inside a bird.

To do Bb's versatility full justice, I must mention another way it sometimes propagates itself. Its life is at hazard each time a tick molts, for the hormones that induce molting also inhibit spirochetes' growth. Bb's population dips sharply as each transformation nears; afterward the numbers usually rise again, and Bb can go on to its next host. However, Bb has developed a backup system in case molting hormones thin its ranks or a generation of ticks cannot find hosts and face a population crash. After colonizing a tick's ovaries, Bb may enter the egg cells there and multiply inside them. If those borrelia survive as the eggs develop into larvae, some of the next generation of ticks will hatch out infected. The number of deer ticks born infected with Bb is tiny, but the proportion is larger in related ticks that transmit Bb in other parts of the world. The infected newborns probably act as insurance against reproductive hard times.

Obviously it takes very specialized adaptations for Bb to

emerge unscathed from all these passages between hosts and between life stages of the deer tick. Bb occasionally stumbles into some tick other than a deer tick or even into a fly or a flea, but apparently something always happens to stop it from traveling on to a person. A rabbit tick may take it to a rabbit, and a mouse tick may take it to a mouse, but neither one bites humans. Bb can infect a dog tick, but it dies out before the tick molts to its next stage; perhaps Bb is done in by that tick's molt-inducing hormones. In the northeastern United States, only *I. scapularis* transmits Bb to people; in other places close relatives of the deer tick transmit local strains of Bb to humans with similar exclusiveness.

This neat system of adaptations is fine for Bb, but what about the tick, or for that matter the mouse, deer and Bb's other hosts? Before proceeding to look at what brought humans and Bb crashing into each other's lives, one must ask, does Bb make ticks sick? And if not, why are people less fortunate?

SIXTEEN

Is the Tick Sick?

NO ONE CLAIMS TO KNOW when a tick feels queasy, but scientists confidently call Bb part of the deer tick's natural flora. This is the term for germs that normally inhabit a host without making it sick. It is true that Bb does not give deer ticks the blind staggers or other visible symptoms, but there are hints that it sometimes does subtler damage. One must ask if the same is true of Bb's most important hosts, the white-footed mouse and the white-tailed deer. And while Bb certainly is not part of humans' natural flora, many of the people it infects show only mild symptoms or none. So the term natural flora does not always convey utter harmlessness, and the answer to whether Bb makes a species sick may not be a simple yes or no.

It is easiest to see the role of natural flora in the host species we know best, ourselves. Our flora first appear quite sparingly, in the womb; not many germs can cross the placental barrier, but a few, including Bb and some other spirochetes, sometimes manage it. Our nearly pristine state ends before we draw breath. As the infant travels through the birth canal, it meets hordes of local microbes, and from the moment it emerges it is under constant assault. Germs accost it from the air and water, the mother's milk and breath, and from everything that touches it. Many of these microbes are digested, dissolved or flushed away; most that get past these first lines of defense are destroyed by antibodies or by germ-killing cells in the blood, lymph nodes, liver and spleen.

However, some microbes that once caused illness have become familiar with the human body; that mutual habituation lets them settle in permanently, especially on the skin and in the mouth, nose, colon and genitals. In the course of a lifetime, a thousand or so microbial species become our constant companions; perhaps 150 more arrive, cause illness or silent infection, and disappear. Although much of our natural flora once caused illness, they and we are now commensals, literally diners at a common table.

The advantages to the germ are clear. If it kills its host, it loses its bed, board and vehicle to another host. As one would logically expect, our deadliest human diseases, such as rabies and bubonic plague, are caught accidentally from other species, not transmitted person-to-person; they are dead ends for humans and microbes alike. A human pathogen is better off with an ambulatory host that can transmit it by coughing, sneezing, diarrhea, sex or just touch. As an alternative, the germ can extend its stay in the host, provoking chronic or recurrent symptoms. That is what many spirochetes do, causing long-lasting bouts of illness that may be broken by periods of quiescence—for example, Lyme disease, relapsing fever and syphilis. This is essential to the survival of microbes such as Bb, which must spend months or even years in some of its hosts awaiting transmission.

The advantages to the host are equally clear. If we cannot kill or completely subdue a germ, we do well to gradually slow it from a lethal gallop to a gentle symptomatic stroll. Still better, we may even extract some sort of repayment for housing it. So despite a few notable exceptions, host-parasite relationships tend to evolve from acute illness to mild or chronic disease, and then sometimes to commensalism or symbiosis. Our symbionts are probably some of our oldest natural flora. Among them are intestinal strains of *E. coli* that manufacture vitamin K, an essential human clotting factor, and the vaginal bacilli that create an acidic environment deadly to many bacterial newcomers. Our natural flora also defend us just by existing: there is limited room at any table, and they fill cellular attachment sites that might otherwise be taken by destructive parvenues.

However, even natural flora can sicken a host when their own lives are disrupted. If antibiotics reduce a person's usual flora, their absence may make more space for the fungus *Candida albicans,* a regular resident of our mucosa; it multiplies wildly and causes thrush. Natural flora may also be a problem if they wander into the wrong neighborhood. They have painstakingly adapted to specific tissues, and their being in unfamiliar organs may cause acute disease, as if they had entered new hosts. As long as the polio virus stays in the human intestine, where it usually goes, it is virtually harmless; if it strays to the central nervous system, it causes paralysis and death. Bacteria that are inoffensive in the nose and throat produce fatal meningitis if they reach the brain. The commonest cause of cystitis is a bacterium that lives peacefully in the colon, and the normal staphylococci of the skin can turn lethal if they colonize a wound.

Even many potentially dangerous germs are docile as long as the host stays fit. Countless millions of people carry the tuberculosis bacillus, but few of them will suffer active TB unless they are malnourished, weakened by other diseases, or have impaired immune systems. Anything that weakens our powers of resistance—old age, psychological depression, other infections, some chemotherapies—makes it easier for both foreign and familiar germs to cause illness. In robust people with intact immune systems, Bb and many other microbes often provoke only mild, transient symptoms or none. Routine blood analysis frequently reveals long-term antibodies to the microbes of Lyme disease, Legionnaires' disease and other illnesses we never knew we had. We once suffered subclinical or mild infections and were never the wiser.

As it is with our natural flora, so it is with the tick's. Bb does not visibly slow down the deer tick (which would make it slow indeed), but there are hints that once it did, and perhaps still may. If tick tissue is cultured in a laboratory and infected with Bb, cellular damage sometimes appears; this may result from living in the lab, but it may occur in natural conditions as well. And if a tick's ovaries are very heavily saturated with Bb, the eggs often fail to develop and may even die. In the tick, as in people, health problems arise when a long-standing commensal such as Bb passes a certain population threshold or stresses a particular organ.

It is not surprising to learn that Bb's chief reservoir and host, the white-footed mouse and white-tailed deer, have also reached a high degree of mutual tolerance with the germ. So have some other small mammals, such as rabbits and hamsters. However, both wild and laboratory rats infected with Bb develop pitifully swollen joints, the arthritis of Lyme borreliosis. And as farmers and pet owners know, Bb can sicken many animals, producing some of the same symptoms as human Lyme disease. Dogs and cats suffer pain, lameness, lethargy and weight loss. If the disease is chronic, there may be complications affecting their eyes, heart, kidneys or brain, leading to seizures or paralysis.

In the Northeast and upper Midwest, Bb and several related borrelia have become serious problems in stables and dairy barns; infected cows and horses show some of the same symptoms as people, though often in milder form. If Bb reaches a cow's placenta, it can cause abortion. Also, Bb is present in infected cows' milk, but fortunately this is no threat to humans, since any spirochetes not killed by pasteurization are digested if we swallow them. To infect a mammal, Bb must travel directly from the tick's hemolymph to the mammal's blood.

For us to tolerate Bb as well as ticks and mice do, we would probably have to live with it through thousands of years of coevolution. Until then, we will have to depend on vaccines, antibiotics and, above all, prevention. It is only in the last century or so that we unwittingly invited Bb liberally into our environment and our bodies. Its effects on people were not reported until the late nineteenth century, and then only in a sporadic, fragmentary way. A mere twenty-five years have passed since Lyme disease was defined, and twenty since Bb was discovered. It was in this evolutionary blink of an eye that humans met a new epidemic, and Bb found a new lease on life.

SEVENTEEN

Rash Discoveries

BB MIGHT HAVE STAYED undiscovered for years longer were it not for two mothers in Lyme, Connecticut, who acted with uncommon intelligence and persistence. In 1975 Polly Murray and Judith Mensch, each unknown to the other, had a child who was mysteriously ill, suffering recurrent bouts of fever, aches and swollen joints. The cases puzzled local doctors, who finally made diagnoses of juvenile rheumatoid arthritis (JRA), an uncommon, noninfectious disease in which a person's immune system attacks his own joints as if they were foreign invaders. Each mother independently concluded that the doctors were wrong and stubbornly pursued a better explanation.

Mrs. Murray, an artist and mother of four, had reasons for doubt. Every member of her family had been sick many times in their fifteen years in Lyme, often with symptoms like those of the son said to have JRA. Their problems had included fever, swollen glands, swollen knees, stiff necks, rashes, sore throats, headaches, eye troubles, facial tics, fatigue and depression. Mrs. Murray herself had been ill almost constantly; she later said she felt as if a war was going on in her body, moving from place to place. Even the family dog was sick, its joints so swollen it was lame, and its legs twitching uncontrollably. Over the years, the Murrays had received diagnoses from unspecified viral infections to lupus to psychosomatic distress. None of these labels was ever confirmed, and no treatment gave lasting help. Adding insult to illness, a number of doctors treated Mrs. Murray like a hypochondriac and a pest for insisting that the family's ills had a pattern, and that their health depended on finding it.

Mrs. Mensch equally doubted her daughter's diagnosis of JRA, and like Mrs. Murray, she had been hearing stories about local people having similar problems. Both women began talking to other parents. They found that their children's symptoms had struck scores of people in and around Lyme, not all of them young, and that almost a dozen had been diagnosed with JRA. By now the two mothers had learned, as smart patients eventually do, that they must become the ultimate experts on their own health. The more they read and talked about JRA, the surer they were that in Lyme it had become a wastebasket diagnosis for puzzling cases. Dozens of cases of JRA in their corner of Connecticut would mean an incidence thousands of times above normal. Also, since JRA is not infectious, it does not occur in clusters, yet in Lyme entire households had become feverish, arthritic and exhausted. In some of them pet dogs and horses were also feeble and lame. There was only one logical conclusion: people and animals alike had been exposed repeatedly to an unidentified poison or microbe.

Already dismissed by some doctors as hysterical and insubordinate, the two women made desperate calls to the federal government's Centers for Disease Control, the state department of health, and the medical school at Yale University. Despite a few initial rebuffs, they were finally taken seriously; to Mrs. Murray's surprise and relief, she was invited to travel to New Haven and marshal her facts before a medical committee at Yale. After she did so, the assembled doctors agreed with her arguments and shared her concern: apparently some strange new disease was appearing in Lyme.

The task of defining the outbreak and locating its source fell to a research team led by Dr. Alan Steere, a rheumatologist with a background in epidemiology. They canvased the adjoining towns of Lyme, Old Lyme and East Haddam, looking for people who had recently suffered swollen joints. They found and interviewed dozens, many but not all of them children; most had also experienced the fever, muscle aches and other symptoms described by Mrs. Murray and Mrs. Mensch. Many had suffered the same symptoms several times, over years and decades. And although whole families had been sick, the illness apparently did not pass directly from person to person. Mapping the cases, Steere saw that they were heavily clustered in and near woodlands;

in rural areas as many as one child in ten might be affected, but almost no cases occurred in the centers of towns or right on the shore of Long Island Sound. Steere also learned that most patients had fallen ill in summer or early fall, and one in four recalled being bitten by ticks.

Taken together, these clues pointed to a microbe rather than a poison or pollutant, probably one from an animal reservoir in the woodlands. At first Steere's group suspected it was an unknown summer virus, one of many that are passed to people by biting insects. However, they noticed that the epidemic was much worse on the east side of the Connecticut River, the side on which deer ticks were abundant. It was already known that deer ticks were spreading eastward across Connecticut and might have a role in transmitting Rocky Mountain spotted fever and the parasitic disease babesiosis. The researchers gave out glass tubes to people in the Lyme area, along with informational letters asking them to collect and send in any insects they thought had bitten them, and to watch especially for ticks.

Early in 1977, Steere's team published their first results. In and around the three towns near the mouth of the Connecticut River they had found thirty-nine children and twelve adults with what was apparently a new disease, perhaps spread by biting arthropods. Steere named it Lyme arthritis, for the town where it was first reported. Some months later, in a follow-up paper, he confirmed the link of Lyme arthritis with severe fatigue, neurological symptoms, and a rash known in northern Europe and called erythema migrans (EM), the wandering rash. A visiting European doctor had mentioned the rash to Steere, and later it would turn out to have greater significance than the researchers had first thought.

That summer, Lyme arthritis caught the attention of the news media. It could have been invented for them, a mysterious, crippling new disease that struck children in conspicuously affluent towns, discovered because mothers had persisted in the face of medical scorn and dismissal. The timing was as good as the story; after decades of overconfidence about defeating infectious diseases, the public had been sensitized by frightening new ones. Lassa fever, Marburg disease and other lethal hemorrhagic fevers had made their debuts. In the summer of 1976, America had fearfully awaited the outbreak of a dangerous new variety of flu virus. The flu epidemic never happened, but the vaccine used against it caused a small epidemic of side effects, enough to scare some people away from flu shots for years. Instead of deadly swine flu, July brought a mysterious pneumonia that struck hundreds of American Legion members celebrating the nation's bicentennial in Philadelphia. Soon they were dying by the dozens. The Legionnaires' disease remained a front-page story for half a year as epidemiologists eliminated false leads, from nickel carbonyl poisoning and bird-borne microbes to radiation leakage and biological warfare by terrorists. Finally they discovered a new genus of bacteria, dubbed *Legionella,* that managed to eke out a living in such ungenerous environments as water towers and air conditioners.

So in the middle of 1977, when Lyme arthritis gave the press another mystery disease, they ran with it. The story broke late in May in the *New York Times* and on CBS television; within days reporters and camera crews descended on Dr. Steere, Mrs. Murray and the less than delighted town of Lyme. All around the country, some people became wary and others downright hysterical when they heard accounts of children on crutches and adults with chronic fatigue, memory loss, pacemakers and facial-nerve paralysis. All the publicity made new reports of Lyme arthritis come pouring in to doctors, public health officials and the press. The media blitz was not only journalism's scare story du jour but the first round of news about what would later be revealed as an international epidemic.

The list of symptoms associated with Lyme arthritis kept growing. In 1979 the usually cautious Dr. Steere published another paper, saying he believed that the syndrome he had described was much more than arthritis. To its formal definition he added the EM rash and several complications of the heart and nervous system, and he proposed changing the name to Lyme disease. By then thousands of cases had been reported in a growing area that included not just Connecticut but Rhode Island, Long Island and Wisconsin. Still, no one knew what microbe caused it or how it was transmitted.

EIGHTEEN

The Magic of Names

THANKS TO LYME DISEASE, the name of a placid, prosperous and otherwise unnoteworthy little town has entered the English language.* With remarkable speed it entered a score of other languages as well, from Latvian to Japanese. In less than a decade after Steere defined and named it, Lyme disease was discovered in Europe, Africa and China, and everywhere its name was Lyme (or a phonetic equivalent) coupled with the word for disease or for borreliosis. In many places, Lyme was just a foreign sound with no particular associations. But in Lyme it meant home, and to some people there the town's sudden celebrity was like a poke in the eye.

True, some people were deeply grateful that an epidemic

---

* Lyme is not just an old spelling of Lime. Lovers of etymological tangles should look up both spellings in the *Oxford English Dictionary.* There they will find *lyme grass, lyme* (or *lyam*) *hounds,* and *limefingered,* but not *Lime House* or *limehouse,* which must be sought elsewhere.

that threatened their families was now identified and might become treatable. Others, especially during the initial media frenzy, were not happy. Mrs. Murray, widely credited with putting Lyme on world maps, received both public thanks and unsigned hate mail. Some of her neighbors feared that if Lyme was widely synonymous with ticks, germs and epidemic disease, its economic development would stall and real estate values collapse. Others complained that friends and relatives who lived in distant cities were now afraid to visit them. A few even claimed that Lyme arthritis was a fiction, a product of mass hysteria. But what apparently inflamed people most was damaged pride and an intuitive shudder before the power of words. They did not want the town's name to forever evoke fear and revulsion.

They were right, of course, about the magic of names. W. H. Auden said that a poet does not skip the begats in the Old Testament or the catalogue of ships in the *Iliad,* he savors them. There is similar poetry in the names of many microbes. A poet could take delight in a Whitmanesque roster of microbes named for the places where they were discovered—Pontiac, Coxsackie and Four Corners (for a city in Michigan, a town in upstate New York and the juncture of New Mexico, Arizona, Colorado and Utah). A list of evocative bacterial names could start with genera named for great figures in medical science—not only *Pasteurella* but *Rickettsia,* which sounds like a kind of creaky furniture but honors brilliant young Howard Ricketts, who discovered and was killed by that group of germs. Or *Listeria,* a word better suited to fragrant blossoms than to a contaminant of soft cheese, named for the great surgeon Joseph Lister.

If folks in Pontiac and Coxsackie are put out because their towns evoke germs, their complaints remain murmurous. But admittedly people do not ask for that kind of fame; no town clamors for a rash or a palsy to trumpet its name around the world. When the deadly new pneumonia of 1976 was called Legionnaires' disease, some Legionnaires did object, but in vain; the term, like the disease, was already known worldwide. I know only one microbial name that yielded to protest, apparently because of the recent form of Grundyism called political correctness. In 1993, when the lethal Four Corners virus was in the news, residents of the Four Corners complained of potential damage to their home, name and pocketbooks. Many were Native Americans, and perhaps in belated deference to their feelings of grievance, the microbe was rechristened. Someone proposed Muerto Canyon (Death Canyon), but that, too, struck the reef of local opinion. Then someone else, with a nice sense of irony, suggested Sin Nombre, which is Spanish for No Name. Others with no Spanish or an equal sense of irony agreed. Sin Nombre it remains.

Place names can evoke angels and devils; just try to ignore the overtones of Auschwitz, Jerusalem and New Harmony. No wonder some people in Lyme hated being paired with an infection that was thought of with fear and loathing. What now seems striking is that the resentment dwindled so quickly. In 1989 a reporter from the *New York Times* asked people in Lyme how they now felt about the town's fame. A few were still put out, but more had come to accept or even enjoy that the place was known around the world. A handful said they were pleased by Lyme's identification with sickness and ticks; they thought it had discouraged tourism and heedless land development, leaving the spacious, affluent town to itself. They were grateful to Bb, which they saw as nature's force for restrictive zoning.

NINETEEN

The Annals of Myopia

WILLY BURGDORFER WAS HELPED to discover Bb by his awareness that Allen Steere had not found a truly unknown disease. Educated in Europe, Burgdorfer knew that for a century various symptoms of Lyme disease had been described there under a dozen other names. A number of researchers had even correctly guessed the disease's cause and vector. The long lack of follow-up to their insights merits an entry in the annals of medical myopia.

Over three decades, from 1883 to 1913, a dozen doctors reported both the early-stage and chronic skin symptoms of Lyme disease. The first was a German named Buchwald, who in 1883 described a skin lesion now known to be a symptom of chronic infection with Bb. Later called acrodermatitis chronica atrophicans (ACA), it was found to be rather common in northern Europe. In 1909 Swedish dermatologist Arvid Afzelius described a patient who developed a spreading, ring-shaped rash after a tick bite—the

bull's-eye rash of early Lyme disease. Afzelius called it erythema migrans (EM), or wandering rash, and guessed the
cause was a microbe or toxin spread by ticks. Four years
later an Austrian named Lipschütz reported that one of his
patients had such a rash for seven months, so he called it
erethyma chronicum migrans (ECM). He, too, thought the
cause was a tick bite, and he said research should focus on
the tick's gut and salivary glands. That would not happen
for another sixty years.

In retrospect, it is astonishing how much evidence piled
up without reaching its natural conclusion. For decades,
doctors in northern Europe linked ACA and EM to one
symptom after another. For instance, in 1922 two French
physicians connected EM with meningitis and nerve inflam-
mations; they said this proved EM was not just a skin condi-
tion but a systemic disease, perhaps caused by a spirochete.
It was not a random guess; in those days, syphilis was often
first recognized by its chancre, or skin sore, and it was
treated and studied chiefly by dermatologists. When dermatologists saw a skin lesion associated with chronic symptoms involving many organs, their first thought was syphilis.

During the 1930s and 1940s, a number of Swedish and
German doctors, still thinking of ACA and EM as different
diseases, noticed that each was sometimes accompanied by
arthritis, heart problems, facial-nerve paralysis and a cluster
of neurological problems called Bannwarth's syndrome.
They kept hearing that these symptoms had followed tick
bites and guessing that the cause was a spirochete. This idea

was bolstered by the discovery that EM patients' blood, tested for syphilis, often gave false positives; it followed that a close relative of *T. pallidum* was present. In the 1940s, when penicillin was seen to cure syphilis, doctors tried it against EM and ACA, and the patients were cured of all their symptoms. Since antibiotics attack bacteria but not viruses and protozoa, this further supported the bacterial nature of EM and ACA.

Around midcentury, several Swedes came close to isolating Bb and proving that it caused EM. In 1948 Carl Lennhoff found spirochetes in the skin of EM patients; his report was first disputed and then ignored. The next year, Sven Hellstrom visited the United States and lectured doctors in Cincinnati about "ECM with meningitis"; he said it was probably caused by a tick-borne spirochete and could be cured with antibiotics. One might wonder why his audience did not dash out, collect ticks, check them for spirochetes, and solve a medical mystery. In fairness, one must say they had reasons. First, collecting and dissecting ticks is specialized, laborious work few doctors are trained to do. Second, no spirochete had yet been found inside any hard tick. And third, neither EM nor ACA had ever been seen in North America. It was the Swedes' problem.

This does not explain why researchers failed to make the definitive leap in Europe, where EM and ACA were fairly common. By the 1950s it was clear to many people that one microbe lay behind EM, ACA and their accompaniment of arthritis, heart problems, Bell's palsy, mood changes and depression. In 1955, in Germany, Hans Götz emulated the heroes of early bacteriology by transplanting skin from ACA patients to himself and other healthy volunteers. All developed EM and arthritis, and all were cured by penicillin. Still, no one proved the links between EM, ACA, ticks and a specific germ.

In 1970 the first case of EM was reported in the United States, by a Wisconsin dermatologist named Rudoph Scrimenti. His patient was a local doctor who had been bitten by ticks while grouse hunting and then developed a rash, arthritis and neurological symptoms. Four more cases were reported in Connecticut in 1975, the year Polly Murray and Judith Mensch started pursuing health authorities about their children's illness. That is when the story of Bb's discovery is usually said to begin, but a century of research on Lyme disease already existed. It and Bb had almost been discovered many times over many decades. Now it seems puzzling that the discovery did not come sooner.

But it did not, and who knows why? Some writers love to declare that a discovery was inevitable. "Given the circumstances," they will say, "it was inevitable that in 1943 someone in West Fludge would explain contagious strabismus." Or, "It was inevitable that soon someone in Hochkatzenspring would solve the mystery of feline hyperkinesis." It sounds portentous and it cannot be disproved, but no time or place really commands discovery. Many eurekas, large and small, follow no logical script. It was very likely that by 1960 someone in Germany or Sweden would discover Bb and Lyme disease, but no one did. It was not inevitable that Willy Burgdorfer would discover Bb twenty years later in Hamilton, Montana, but he did—as he himself has said, by serendipity. We are certain only that his chance of making the discovery was enhanced by knowing about EM and ACA, which few Americans did. Otherwise Bb might have been discovered later, by someone else, and inevitably this book's title would have been about a germ with a different second initial.

TWENTY

From Bitterroot to Lyme

WILLY BURGDORFER SAYS he discovered Bb by serendipity, which in the original sense of the term is quite true. The word serendipity comes from Horace Walpole's tale *The Three Princes of Serendip*—not a fictional place but an ancient name for Ceylon. Walpole's princes repeatedly stepped in muck and came up with gold; or as Walpole more elegantly put it, they were "always making discoveries, by accidents and sagacity, of things they were not in quest of." Today serendipity more often implies chance than sagacity, but Burgdorfer's discovery required both. It was the sort of accident that confirms Pasteur's dictum: Chance favors a prepared mind.

In his native Switzerland, Burgdorfer was an expert on ticks and tick-borne relapsing fever. In the 1950s, when he came to the United States and began working for the National Institutes of Health, that disease was even rarer here than in Europe. He took up research on a different tick-borne disease, one peculiar to his adoptive home, Rocky Mountain spotted fever (RMSF). That deadly infection was on the wane, but soon it would have a frightening resurgence, under circumstances similar to those that caused the spread of Lyme disease. It was the study of RMSF that led Burgdorfer to Bb.

Like Lyme disease, RMSF began as a puzzling local outbreak. It first appeared in western Montana, where the Bitterroot River carves a long valley in the Rockies. Settlers arrived in the 1840s and began felling trees on the rugged hills west of the river; on the more inviting eastern side they built towns, apple farms and sheep ranches. By the 1880s the western side was almost deforested and reduced to scrub, which made it decent country for mining and grazing but superb country for rodents and ticks. Isolated cases began to appear of a lethal summer fever. At first people called it black measles or mountain fever, but the name that stuck was spotted fever. The afflicted developed a soaring temperature, delirium and a distinctive rash that completely covered the body, even the palms of the hands and soles of the feet. There was no mistaking it for anything else, and neither settlers nor the native Salish Indians had seen anything like it. It especially struck young men and boys, and in some years it killed up to 90 percent of its victims.

For two decades the number of cases kept growing, and by the early 1900s the Bitterroot Valley's farmers and townspeople were near panic. News of the deadly epidemic spread, and it caught the interest of several researchers, among them a brilliant young pathologist in Chicago named Howard Ricketts. He went to Montana and soon focused his research on ticks; it was an act of intellectual daring at a time when the notion that arthropods carried disease was newer and more challenging than the idea that germs caused it. Only in quite recent years had mosquitoes been shown to transmit malaria to humans, and ticks to carry the babesia parasite of Texas cattle fever. Soon research would prove that arthropods also carry the germs of typhus, yellow fever, sleeping sickness and bubonic plague. Later, vector control would reduce the incidence of mosquito-borne diseases and virtually wipe out tick-borne cattle fever. But in 1906, when Ricketts set out to find the cause of RMSF, no tick had ever been shown to carry bacteria or to transmit any disease to humans.

Within a few years, Ricketts discovered the RMSF germ, a tiny, round, elusive bacterium unlike any seen before. It was just the first of a whole family of microbes, called rickettsia for their discoverer, transmitted by arthropods and responsible for a number of major and minor diseases, including typhus. Ricketts went on to prove that the germ of RMFS, *Rickettsia rickettsii,* is carried by the Rocky Mountain wood tick, from a natural reservoir in wild rodents. He dissected wood ticks, described their reproductive cycle, and discovered the germ's transovarian passage from one generation of ticks to the next (a more important reproductive path for *R. rickettsii* than for Bb). After launching research on an RMSF vaccine, Ricketts left for Mexico to study typhus, and there in 1910, when he was only thirty-nine, it killed him.

Some forty years later, when Burgdorfer arrived in the United states, RMSF had become rare, though now it was known to occur throughout the Western Hemisphere, carried by different ticks in different regions. The disease could be minimized by avoiding ticks, blocked by a vaccine, and cured by antibiotics. In its original home, the Bitterroot Valley, it had almost vanished, because the conditions that gave rise to it had changed. People had always been accidental hosts to the Rocky Mountain wood tick and to *R. rickettsii*. RMSF began as a disease of loggers, ranch hands and children, who became infected as they worked and played in the valley when deforestation made it a virtual nursery for ticks. As the valley kept developing, towns and fields replaced scrub; the ticks' habitat dwindled, and with it rickettsia and RMSF. Still, without diagnosis and treatment spotted fever could kill, so nationwide surveillance continued, and a handful of researchers continued studying RMSF. Burgdorfer was one of those few, at the federal government's Rocky Mountain Laboratory in Hamilton, Montana.

The caution about RMSF was justified. It returned after World War II, especially in the Northeast; by the 1970s there were dozens of cases each year in Connecticut and Long Island, some of them fatal. Burgdorfer had predicted this would happen, because RMSF can spring up wherever people create the scrub and patchy woods that favor rodents and ticks. The postwar explosion of suburbs outside such cities as Cincinnati, New York and Washington, D.C., had done just that. And as Burgdorfer had also predicted, RMSF peaked and declined during the 1980s, as continuing development paved over newly created suburbs.

In 1981, at the height of the scare over RMSF's reemergence, Burgdorfer was seeking the source of the epidemic in the Northeast. No one knew which tick carried *R. rickettsii* there, but Burgdorfer's first guess was the dog tick, the germ's most common vector outside the Rockies. Public health workers in Long Island collected several thousand dog ticks and sent them to Burgdorfer's lab in Montana. He dissected them, looking for *R. rickettsii,* but came up empty. Then he wondered whether the germ's local host might be the deer tick, now the most common tick on Long Island. Several hundred deer ticks were captured and sent to him, but again he found no RMSF microbes. Then several hundred more were sent from Shelter Island, off the coast of Long Island, where some cases had recently appeared.

Following his usual routine, Burgdorfer dissected the ticks with an eye surgeon's scalpel, stained their hemolymph, and looked for rickettsia under the microscope. Again there were no rickettsia, but in two ticks he found microfilariae, the microscopic form of a parasitic worm. Wondering if he would find more, and if they were somehow significant, he continued to dissect the two ticks tissue by tissue. There were no more worms, but in a smear made from a tick's midgut, he saw large, irregularly coiled spirochetes spiraling lazily. No spirochete had ever been found in a hard tick, and these were not quite like any he recognized. He dissected another 125 of the ticks from Shelter Island and found the spirochetes in two-thirds of them.

Then a bell rang in Burgdorfer's memory. He recalled reading papers about EM, the European rash-cum-arthritis that had raised suspicions of a tick-borne spirochete. He thought especially of Lennhoff's paper and of Hellstrom's lecture in Cincinnati thirty years earlier. The symptoms of EM resembled those of Lyme disease; both conditions might be carried by ticks, and both often yielded to penicillin. Perhaps, thought Burgdorfer, EM and Lyme disease are the same thing, caused by this spirochete from Shelter Island. Like so many earlier guesses about EM, ticks and spirochetes, this might have led nowhere. But Burgdorfer had a great advantage over his predecessors: he could grow the finicky spirochete in his lab and experiment with it. In the course of research on relapsing fever, his colleague Alan Barbour had recently devised an artificial growth medium, containing some fifty special additives, in which *Borrelia* could thrive in vitro.

Burgdorfer cultured the spirochetes from Shelter Island and set out to test his hunch. Injected into lab animals, the germs caused Lyme disease symptoms. He obtained blood samples from convalescing Lyme disease patients on Shelter Island; tests showed antibodies proving exposure to the same spirochete he had found in local ticks. Now Burgdorfer was convinced that he had discovered the cause of EM and Lyme disease, which were one and the same disease. It was a previously unknown species of *Borrelia,* the twenty-first to be discovered, and it lived in the all-too-common deer tick.

Burgdorfer published his results in 1982, and the spirochete was named in his honor. During the next couple of years, one stroke of confirmation followed another. Researchers found Bb in the skin, blood and cerebrospinal fluid of Americans with Lyme disease and Europeans diagnosed with EM, ACA and Bannwarth's syndrome. By 1985 Burgdorfer had shown that ticks carrying Bb existed in much of the United States; in Europe closely related ticks carried other strains of the same germ. Before the decade was over, Lyme disease and ticks bearing Bb had been found in Russia, North Africa, China, Japan and Australia. Which raised the question of when and how it got to all those places.

TWENTY-ONE

Far from Primeval

IT IS UNDERSTANDABLE that people failed for a while to identify Lyme disease in all its complexity, but the EM part, the bull's-eye rash of its first phase, is hard to miss. If Americans did not notice it before the 1970s, perhaps it wasn't here. One must ask, when did Bb get to Lyme? Was it an American native, long unnoticed, or a recent migrant from some other ecosystem? If a long-quiescent native, why was it suddenly visible? Or if a migrant, how did it arrive, and from where? Similar questions can be asked about Bb's emergence in Europe. Did it smolder there before its effects were noticed in the 1880s or was it introduced then from somewhere else? Did Europe take it to Africa and Asia or did one of those continents export Bb globally?

We cannot answer many of these questions yet, and some of them may be too big for mature minds. Immature minds love big questions, the ultimate whys of things. Freshmen are fond of grappling with nothing less than the nature of truth and the fate of civilization, and their answers are often predictably gassy. One of the great pleasures of teaching is watching people learn to ask more modest, answerable questions. They progress from broad queries about *why* to specific ones about *how*. Having settled for grasping one small corner of the fabric of life, they find, paradoxically, that they get a better sense of its larger patterns. So in the spirit of mature inquiry, and to show that I know how to follow my own advice, I will narrow the focus of this chapter to Bb's history in one small, well-observed place—Lyme and its neighboring towns, where Bb was discovered.

The story of Bb's life in Lyme turns on the character and use of the land. The land is the Connecticut River Valley where it joins Long Island Sound, and its character and use have changed many times. When you look at the sand, salt marshes and wooded hills that now surround Lyme, it tickles the mind to think that once this place snugged against the hump of Africa, near Casablanca or Tangier. Later, when Africa and the Americas had drifted apart, ice sheets periodically crawled over this land and retreated; glaciers and forest succeeded each other many times over. After the last glacier withdrew, about 20,000 years ago, Connecticut was covered by what would later be called forest primeval—a phrase by Longfellow, who liked to imagine that his forebears had discovered unspoiled peoples in an unspoiled Eden.

Actually, when Europeans arrived, most American forests had not been truly primeval for more than 10,000 years, since Asian hunters crossed the land bridge that once joined Siberia and Alaska. As they spread through the hemisphere, they hewed trees and set fires, first to catch game and then to clear land for planting. Later they changed the environment more drastically, when they took up large-scale farming. In the Midwest and Mesoamerica, they created fields of maize and beans so vast that they supported cities the size of any then in Europe.

Incidentally, exploiting the land this way did not bring them health and plenty. They suffered the declining vigor that usually strikes people who have turned from the hunter-gatherer life to farming. They caught viruses from their herds, worms from their pets, and bacteria from scavenging birds and rodents. Epidemics brewed in their middens and privies. Reliant on high-carbohydrate staples, they became less robust than their ancestors; the remains of pre-Columbian farmers reveal deficiency diseases, multiple infections, shortened stature and shortened lives. Few of them reached middle age. Some of the tribes Europeans met were mere remnants of empires that had exhausted their land and crashed. Their fields had reverted to second-growth forest, and Europeans who had never seen unexploited land mistook it for virgin country.

In New England, the natives' farming had stayed on a small scale, so the population remained modest. The forest there was not quite pristine, but enough dense, mature woodland remained to make William Bradford, fresh off the *Mayflower,* describe Massachusetts in 1620 as "a hideous and desolate wilderness, full of wild beasts and wild men." Soon his fellow Puritans moved west to the gentler, more inviting Connecticut River Valley. In 1635 a small band of them settled near the river's mouth, on its west bank; their little fort there became the village of Old Saybrook. A few decades later, their children were building villages across the river, and these became Old Lyme, South Lyme and Had-lyme.

These transplanted Puritans in time became Connecticut Yankees. Some made their living from the sea, as fishermen, whalers and shipwrights; others felled trees to create a patchwork of small farms. These farmers were not the only force changing the ecosystem. Transatlantic traffic was mixing Old World and New World biotas, in a rather one-sided exchange. America received the sparrow and "Kentucky" bluegrass from England, tumbleweed from southern Russia, and a mob of microbes (causing smallpox, measles, typhus and more) from Europe and Africa. In return, the Americas exported tobacco, potatoes and the potato blight that started Ireland's great famine. The exchange continues today, as imported zebra mussels clog Northeastern waterways, the Korean hantavirus invades Baltimore, and Asian mudfish waddle down Florida's roads.

Some scientists think Bb was in Lyme long before Europeans arrived. Perhaps, they say, it came from Asia 10,000 years earlier, in ticks clinging to wildlife that crossed the Arctic land bridge, or in ticks on migrating birds. They point to genetic differences among *B. burgdorferi* strains in the East,

Midwest and West Coast, which suggest that it underwent long local evolution in each of these regions. However, others think Bb arrived with early Europeans, in a tick attached to a ship rat or an imported sheep (sheep ticks are Bb's main vector in Europe). Bb could have come that way directly from Europe or, through European trade, from Asia or Africa. In fact, over the past few centuries it could have traveled in ships from almost any place in the world to any other.

None of these theories has much factual support yet, and some people doubt that Bb has been in Lyme for long. They say it probably emigrated during the past century, and decline to even guess where from. If it was there in the Pilgrims' day, they ask, why did it take 350 years for people to notice such egregious symptoms as a bull's-eye rash and childhood arthritis? In reply, those who consider Bb an old American argue that Lyme disease probably did occur sporadically, but it went unrecorded by people more concerned about frequent epidemics of smallpox and diphtheria. They add that European settlement, by altering the landscape, may have first increased Lyme disease and then made it almost vanish, as RMSF did in the Bitterroot Valley. If they are right, Bb's modern appearance is a second coming. Indeed, there is patchy evidence that something like that may have happened, at one point bringing Bb near extinction in the Northeast.

In 1749 the Swedish botanist Peter Kalm, a friend of Linnaeus, kept a detailed diary as he traveled through the American countryside. One warm June day, in upstate New

York, he noted in exasperation that the woods there "abound with woodlice [ticks], which were extremely troublesome . . . scarcely any of us sat down but a whole army of them creep upon his clothes." Later, in his *Travels in North America*, he wrote, "This small, vile creature may, in the future, cause the inhabitants of this land great damage unless a method is discovered which will prevent it from increasing at such a shocking rate." It is not certain today which species of tick plagued Kalm's party, but they may well have been deer ticks.

The tick problem solved itself. A half-century later, most of the Northeast's forests had been razed for timber, fuel and farmland. In 1804 Jedidiah Morse, a Connecticut native who became America's first eminent geographer, observed that his home state was now entirely crisscrossed with roads and dotted with houses and barns. "The whole state," he said, "resembles a well cultivated garden." A tidy garden, of course, is the last thing ticks can stand. In 1872 naturalist Asa Fitch retraced Peter Kalm's footsteps and saw that the woodlands he had described were gone; as a result, said Fitch, the ticks were "nearly or quite extinct."

One reason for the scarcity of ticks was that as forests vanished, so did deer. In William Bradford's day, the United States had about 25 million deer; in Fitch's time fewer than half a million remained, and in some eastern states they had vanished entirely. Other deer tick hosts, such as raccoons and foxes, were also scarce. Thoreau wrote in 1854 that deer had not been seen near Walden Pond for eighty years;

the biggest animal left alive there was the muskrat! If Bb really was around in the colonial Northeast, it later got squeezed into a very tight corner as its habitat and hosts nearly vanished. Deer and their ticks survived mostly in small herds on Nantucket, Shelter Island and other havens off the coasts of New York and New England. Yet even as people predicted their extinction, the land and its use were changing again.

In the mid-nineteenth century, New England's soil was crowded and tired, but a land bonanza had opened up in the Midwest and the Great Plains. A steady stream of New Englanders abandoned small farms to go west; many who stayed also left the land, taking jobs in mills and factories. As the Northeast's farms shrank in size and number, unused fields returned to forest; between 1860 and 1890, woodlands there increased fourfold. But this was not the dense forest the Pilgrims thought so wild, it was sparser second growth. Connecticut, once unbroken forest and then unbroken farms, was turning into a patchwork of villages, fields, woods and scrub—an ideal environment for deer and ticks.

With a strong push from Theodore Roosevelt and the nascent conservation movement, federal, state and local governments began acting to protect woods and wildlife. Deer populations started rebounding in the early twentieth century, but predators such as wolves, bears and big cats did not. The recuperating land also had a resurgence of birds and small mammals. In 1938, when half of Connecticut was forest again, the WPA guide to the state* said that no place in America offered better rabbit hunting, and again there were raccoons, foxes, muskrats, mink and otter. "Even the white-tailed deer, dazed by the glare of approaching head-lights, often stands rigid in the center of the less frequented roads."

Some of the abandoned farmland became not parks and nature preserves but sites for new housing. Rehabilitated countryside was attracting people in flight from noisy, crowded cities and shabby mill towns. Urbanites who had never seen true wilderness mistook second- and third-growth woods and neglected pastures for preadamite nature. They were happy with it as it was, because they did not really yearn for the wilds; they just wanted greenery and birdsong with amenities, and perhaps a tree house for their children such as they themselves had never enjoyed. As this new commuter class swelled, suburbia became a major part of American life and the American landscape.

After World War II, bigger families and a new prosperity made suburban growth explosive. Yet despite all the new

---

* *Connecticut: A Guide to Its Roads, Lore, and People* was produced during the Great Depression by one of Franklin Roosevelt's brightest brainchildren, the Works Progress Administration. The WPA hired hundreds of unemployed writers to produce the American Guide Series, a volume for each of the forty-eight states. Exhaustive and well written, the WPA guides are still unrivaled compendia of geography, local history and folklore, and good companions for driving America's byways. Now they also offer an intriguing window on America in the 1930s.

houses, roads and malls, the net amount of greenery was increasing. More land went into parks and preserves, and more abandoned farmland became pockets of semirural housing. Although today Connecticut does not have one acre of its original hardwood forests, two-thirds of it is young woodland, and one-sixth of it farms. There is also more ecotone than ever, "edgy" landscape that is transitional between woods and clearing—the favorite home of deer and ticks. Furthermore, many suburbanites have planted their yards with just the sort of shrubs and young trees deer prefer.

Today there are as many deer in America as when Bradford landed, and some people in the Northeast think they have more than their share of wildlife. Coyotes and moose once again roam Massachusetts, even the suburbs of Boston. Deer graze in Connecticut's suburban yards and amble through parks in Philadelphia. No longer shy, they have adapted to man-made environments as comfortably as squirrels and raccoons. Many Long Island and New England homeowners think of them not as Bambis but as vermin; they call them "rats with hooves." Along with the deer, smaller mammals have returned, and with them their ticks. *B. burgdorferi* probably has a better home in the Northeast than it ever did in the past. That is certainly true in Lyme, Connecticut.

The town of Lyme faces a narrow shelf of sand and salt marshes; behind it rise the wooded hills that line the broad Connecticut River as it nears Long Island Sound. Lyme is a trimly pretty town of green and white, with elm-shaded streets, white–spired churches, tidy Colonial-era houses, and many homes that stand on an acre or more of yard and garden. From Colonial times until the late nineteenth century, most of the surrounding land was divided into small farms that raised hay, grain and livestock. The towns along this coast were once bustling little maritime centers where whalers and fishermen worked, and vessels unloaded Cuban rum and elephant tusks to be made into piano keys. But by a century ago, its old ways were fading. The Connecticut River's salmon had long been fished out, but shad still ran every spring, and Lyme was known for summertime shad dinners in country inns.

As fishing and farming declined, tourism started to take up the economic slack. Vacationers came by the day or week to enjoy the shore; Lyme and Old Lyme especially drew artists and writers. Wealthier newcomers bought summer homes there, perhaps not realizing that the quaintness and quiet they relished were the unnatural stillness caused by vanished jobs and scattered families. Around World War I, there were still some cow pastures and shad fishermen, but barns were being replaced by country homes, and fishing docks by marinas. After World War II, Lyme attracted a new crop of year-round residents, many of them commuters to New York or Hartford. Despite all these changes, there was still a rather country-like atmosphere, and children could play barefoot in the yards and woods.

It was in the decades after World War II that people

noticed ticks were becoming more common. They had reached the mainland on deer returning from offshore refuges. The deer fed on brush and saplings that grew in abandoned farms and spacious yards. When Polly Murray looked at photographs taken of her house in Lyme in the 1920s, she saw sparse vegetation and a few lone trees. In the 1970s, she noted, "the land about our house was thick with trees, wild roses, barberry, bittersweet, wild grapevines, and other vegetation." In the summer she and her neighbors might pick a dozen ticks a day off their children and pets— not the big engorged dog ticks they used to see but tiny "seed" ticks, the nymphal stage of *I. scapularis*.

In the 1980s deer ticks and Lyme disease were spreading inland from the coast, up the valleys of the Connecticut and Hudson rivers, and west from Cape Cod in Massachusetts. Epidemiologists wondered whether this was Bb's first appearance in those areas or a return to old haunts. There was anecdotal evidence that Bb had been around in the past; for decades people in Montauk, at the eastern tip of Long Island, had spoken of "Montauk knee" and "Montauk spider bite"; these were probably manifestations of Lyme disease. Soon it became possible, through new laboratory techniques, to identify very small amounts of Bb's DNA or of antibodies against it. Scientists began looking for Bb's chemical footprints in specimens preserved in alcohol in museum collections. Bb's DNA turned up in ticks collected in Long Island and from islands off New England in the 1940s. Then they were seen in ticks collected in South

Carolina and Florida between the world wars. Molecular traces of Bb were eventually found in the bodies of white-footed mice collected in Massachusetts in 1894. In Europe, researchers discovered evidence of Bb in a fox caught in Austria in 1888 and a cat preserved in Hungary in 1884.

The histories of Bb and Lyme disease are just starting to emerge. Plainly both are at least a century old in North America and Europe. Perhaps, as some people suggest, Bb mutated a century ago from a fairly benign state to a more infectious or virulent form, so that Lyme disease became more common, more severe or both. But even if that is true, Lyme disease could not have become epidemic until a changing ecosystem revived Bb and its hosts, and suburbanization thrust people into their midst. Another fact favoring this environmental theory is that Bb and Lyme disease are by no means the only such case.

TWENTY-TWO

Machupo and Other Disturbances

BB IS JUST ONE OF THREE GERMS now known to be transmitted to humans by deer ticks. And it is only one among dozens that have been causing epidemics as people change their lives and surroundings. Several besides Bb could be paradigms for emerging epidemics; two with typical and interesting stories are the Machupo and Junin viruses. Tiny, round viruses with RNA cores and spiky exteriors, they look quite unlike Bb. Furthermore, their homes and means of transmission could not be more different; for instance, the Machupo virus is native to the Amazon lowlands of eastern Bolivia, and it spreads directly to people from mice. But like Bb, this virus attacks humans only when they crash into its ancient ecological web, pursuing better lives. And like Lyme disease, Bolivian hemorrhagic fever results from a biological domino effect. Its deadly eruption in the town of San Joaquín evokes the rhyme about the dog that ate the cat that ate the rat that ate the cheese.

The Machupo virus once lived silently in Bolivia's Beni province, a dry, sun-baked plain broken by patches of forest. Until fifty years ago, Beni resembled parts of the old American West, a dusty, thinly peopled place where cowboys rode herd on stringy beef cattle; the cows were shipped down the Amazon for export, and imported food was shipped back to feed the cowboys. Much of the region was owned by a big Brazilian company until 1952, when a social revolution parceled out the land to native Bolivians. That ended the cattle economy, and with it imported food, so the people turned to subsistence farming. Naturally they plowed not the dry, leached-out prairie but the rich earth of the woodlands. Such independent farming was supposed to bring freedom and plenty; instead it brought sickness and death.

As usual, the ecological web was more complex than anyone knew. On the borders separating Beni's woodlands and plains lived a mouse called *Calomys callosus*. Suddenly surrounded by new farms and gardens, this little scavenger thrived, and like many wild rodents, it reproduced in greater numbers when food was plentiful. Its population soared. And wherever the mouse went, it left little spatters of urine, shedding a virus that gave humans a hemorrhagic fever. By 1960 people in Beni were falling prey to a mysterious infection that led to internal bleeding and shock, and it killed one victim in six.

In 1962 a nasty outbreak hit the town of San Joaquín. As scientists later figured out, it had been triggered by spraying the area with DDT to halt mosquito-borne diseases. The insecticide, having entered insects, became concentrated in the bodies of small lizards that fed on them. From lizards it rose up the food chain to the town's cats, which ate the lizards; it accumulated in the cats' livers and eventually killed them. The cats had been *Calomys callosus*'s only major predator; in their absence, the mice multiplied like wild and left virus-laden urine all over San Joaquín. Thus DDT poisoned the bugs that fed the lizards that killed the cats that spared the mice that passed the virus to people. When hundreds of mousetraps were set in and around San Joaquín, the epidemic waned.

By the mid-1960s, the cause and transmission of Bolivian hemorrhagic fever had been discovered and its agent dubbed Machupo virus. Today health workers still patrol Beni's towns and ranches on horseback, catching mice and checking their bodies for the enlarged spleen that betrays the virus's presence; if mice are infected, a trapping campaign is carried out. As a result, Bolivian hemorrhagic fever is rare, but the microbe lurks in the environment, waiting to break out if rodent control slackens. The San Joaquín epidemic is a good example of how even a small disturbance to an ecosystem is like someone rolling over in a crowded bed; it forces everyone else to move, and it may push someone over the edge.

A similar disease broke out around the same time in Argentina, because of a related germ and a related mouse. Argentine hemorrhagic fever emerged because cheap herbicides had made it possible to kill the pampas' dense native grasses, making room for commercial crops. When tall fields of maize sprang up east of Rio de Janeiro, no one anticipated that they would encourage an understory of grasses that flourished without direct sunlight. And no one knew that these grasses' seeds were the main food of a local Calomys mouse, *C. musculinus,* which quickly changed from a minor scavenger to the ecosystem's most prolific mammal. Its urine contained a relative of Machupo virus, the Junin virus, which similarly spread to humans in aerosol form. A hemorrhagic fever struck farmworkers, killing one in ten. Since the first outbreak, rodent control has kept Argentine hemorrhagic fever from running riot, but it still occurs, and its range is increasing.

Like the Machupo and Junin viruses, the related but deadlier Lassa fever virus of Nigeria spreads directly from rodents to people. Many other germs similarly reach us straight from wild or domestic animals—the influenza virus, for instance, which mutates to new forms each year in Asia's pigs and ducks. HIV may be another; it probably first reached humans from monkeys as a rising African population, changing jungle to farmland, invaded their habitat. However, Bb is one of many new human pathogens that need arthropod vectors, such as mosquitoes and ticks. Fortunately for much of the world, several new tick-borne diseases are so far confined to where they were discovered— Omsk hemorrhagic fever to Siberia, Kyasanur Forest disease to southwestern India, Crimean-Congo fever to Africa and the Near East. Like Rocky Mountain spotted fever and Lyme disease, they emerged when farming, hunting and other human activities exposed people to infected ticks. So Bb and Lyme disease, though in some ways unique, follow a pattern common to infections that rise from perturbed environments.

Lyme disease is not the only infection deer ticks now transmit to people. Another is babesiosis, the first tick-borne disease ever discovered. Babesia are protozoa resembling the *Plasmodium* parasites of malaria; like plasmodia, they invade red blood cells and can damage the liver, kidneys, and brain. In 1893, when Texas cattle fever was wiping out herds around the world, Theobald Smith and F. L. Kilbourne proved that the cause was babesia transmitted by hard ticks—a startling discovery, because few people had thought any germ could survive passing back and forth between ticks and mammals. By using cheap but effective cattle dip, Smith and Kilbourne gave the first example of stopping an epidemic by tick control.

Babesia seemed strictly a veterinary problem until 1957, when a man in Zagreb, in the former Yugoslavia, was made sick by *Babesia divergens,* a tick-borne parasite of cattle. This seemed to be a medical oddity; the microbe had no history of infecting humans, and the patient's immune system had already been weakened by removal of his spleen. However, more such cases appeared in Europe, and in 1969 human babesiosis started turning up on Nantucket Island, off Massachusetts, in people with normal immune systems.

The culprit in New England was *Babesia microti,* which usually infects mice. As research would show, babesia were emerging as human pathogens in Nantucket much as Bb would in Lyme, and they were both being transmitted by the same tick.

Nantucket's landscape, like Lyme's, had changed several times. The first Europeans there found forest, deer and perhaps deer ticks. They hunted the deer to extinction and felled the trees for pasture; by the eighteenth century, most of Nantucket was a lawn cropped close by sheep, quite discouraging to ticks. A century later, the island was changing again; pine trees were planted as windbreaks, and as overgrazing reduced sheep farming, pastures reverted to scrub. During the twentieth century, tourism and the building of summer homes created the same mix of ecotone and semirural living that nurtured Bb. Now the island was more inviting to ticks than ever before. All that was missing for a tick-borne epidemic was deer.

Deer did reappear, though not on their own. The first one arrived by accident in 1922, when a Nantucket fisherman rescued an exhausted buck swimming out from the mainland. Perhaps because rich summer residents wanted the illusion of natural wildlife, more deer were imported from Michigan in the thirties. During the years that followed, the deer population rose to hundreds, then over a thousand. With the imported deer came deer ticks, and soon they supplanted the old dominant local variety, a mouse tick called *Ixodes muris.* The mouse tick had probably been carrying babesia from rodent to rodent without affecting people. By the 1970s *I. muris* had almost vanished, babesia had shifted to the deer ticks that replaced it, and the deer ticks were biting people.

Today in Nantucket and Connecticut, many deer ticks carry both Bb and *Babesia microti*. People can catch both infections from one bite, and together they make a nasty package. Human babesiosis pops up with rising frequency in Europe, New England and the upper Midwest; growing medical awareness may explain some of the increase, but babesiosis, like Lyme disease, is apparently spreading. While reported cases are still few, the true number of infections must be far larger. Babesiosis is usually mild or silent in healthy young people, but it brings down the elderly and immunosuppressed with fever, fatigue and malaria-like chills that last for days or even months. Many symptomatic cases are probably still misdiagnosed.

Some ticks that carry Bb and babesia now also transmit *Ehrlichia,* a genus of rickettsial bacteria that invade white blood cells. They cause symptoms as mild as malaise and low fever or as severe as pneumonia and encephalitis; untreated, ehrlichiosis is sometimes fatal. Like babesia, ehrlichia were first thought limited to animals, and once they probably were. One kind, *E. chaffeensis,* was discovered in 1935 in Algerian dogs and then in other domestic animals in Africa and Europe. Later it was found to infect people as well, causing human monocytic ehrlichiosis (HME). The first American case appeared in Arkansas in

1986, and now hundreds have been seen in more than thirty states. Similarly, *E. equi* was first thought to sicken only horses, but then it was found to cause human granulocytic ehrlichiosis (HGE). In the United States ehrlichia are carried by several ticks, including the dog tick and deer tick. The early stage of ehrlichiosis produces flulike symptoms such as fever, headache and muscle aches, so it may sometimes be confused with Lyme disease. Some people do catch both infections at once, perhaps from a single tick bite.

Clearly one of the most compelling facts about Bb today is the company it keeps. The microbes causing Lyme disease, babesiosis and ehrlichiosis now coexist in ticks and in people. Apparently they do so more frequently, and they have all emerged as human illnesses for the same reasons. We have created a better, safer world for their hosts and their vectors, and at the same time have generously put ourselves in their way, lending our bodies to them as hosts. From the germs' point of view, the changes we have made in their environment and in our own lives are a gift.

TWENTY-THREE

With Apologies of Sorts

THIS BOOK IS MEANT TO SKETCH the germ's life and its place in our shared ecological web. But now, with apologies of sorts to Bb, I must pull back the camera for a wider shot, as it were, and include the human perspective. No matter how interesting Bb is in its own right, I must respect the fact that if it did not cause Lyme disease, we would not know it existed—and if we did know, few would care. To Bb we are a significant accident but a dead-end host; to us the germ is something else, a source of fear and sometimes suffering. In fairness to my own species and our natural preoccupations, I want to look more closely at the human side of a meeting with *Borrelia burgdorferi.*

As we have seen, the encounter begins when an infected tick bites a person and begins to alternately feed and flush out excess fluid. During the first day or two of attachment, it does not cause Lyme disease; more than twenty-four hours must pass before its saliva transmits enough spirochetes to infect a human being. (That is why daily monitoring for ticks is so important in high-risk areas.) Bb does enter the wound in infective numbers on the second or third day; some spirochetes remain in the skin, near the bite, but others are soon carried through the person's body in his blood. This is the best and worst of times for Bb; blood is an excellent nutrient and a good means of travel and transmission, but also a hotbed of immune activity. As Bb rotates lazily in currents of blood, it begins the classic microbial race with its host—Bb's power to spread and multiply versus the immune system's speed and ingenuity.

Most germs start this race a step ahead, and Bb is no exception. Like the white side in chess or the team with possession in a ball game, it has the advantage of making the initial move; its host starts out on the defensive, not acting but reacting. As time goes by, the host will have opportunities to catch up, but Bb can modify its surface proteins to evade or mislead immune mechanisms. Almost always in microbe-human encounters, the germ is quicker to respond and adapt, because of its short reproduction time (minutes or hours compared to a human's twenty years) and its greater tendency to mutate.

Despite Bb's initial advantage, our immune systems have many silent successes for each failure. Many people whom Bb enters manage to kill it quickly, before infection can take hold. Quite a few others become infected but show only mild symptoms or none; they go undiagnosed and elude the statistics on Lyme disease. In parts of New England and New

York State, half the people who carry antibodies against Bb (proof of past infection) cannot recall having had any Lyme disease symptoms. We do not know why one person is infected by tick bites but another is not, nor why certain people, though infected, show no symptoms. The reasons probably lie sometimes in the germ—how virulent a certain strain of Bb is or how well it eludes human defenses—and sometimes in host traits such as age, nutrition or genetic makeup.

When symptoms do develop, they arrive in three stages; however, these may overlap, and any one may be inapparent. The first stage, which starts a few days to a few weeks after infection, is purely local; a rash appears around the tick bite, often in a bull's-eye pattern, and it sometimes lives up to the name erythema migrans by spreading. This rash is seen in more than half of Lyme disease patients and may occur unnoticed in others. Sometimes Lyme disease goes no further; it is stopped by the body's defenses, alone or with help from antibiotics. But in other people the rash persists or recurs, and second-stage symptoms appear, signaling that Bb has traveled throughout the body.

Bb may start to advance past its entry point just a few days after infection occurs, but second-stage symptoms do not appear for several weeks to several months. They begin with flulike reactions such as fever, chills, fatigue, swollen glands, headache and muscle aches. Then come other symptoms, reflecting Bb's unusual ability to inhabit many tissues, from the skin and joints to the heart and brain. Lodging in the skin, Bb makes the rash continue and in some cases spread over the body and limbs. Swimming in the synovial fluid that bathes our joints, the germ sets off arthritis. In the eyes it can cause conjunctivitis and vision problems. It may also infect and inflame heart muscles, making the heartbeat irregular. In the brain and cerebrospinal fluid, it produces signs of meningitis such as stiff neck, headaches, light sensitivity and mental confusion. Like the spirochetes of syphilis and relapsing fever, Bb can cross the placenta and may sometimes harm the fetus.

In nine people out of ten, the immune system catches up with Bb during the first or second stage and eventually overcomes it. In the tenth person, Lyme disease becomes chronic; it can persist for months, years or even a decade. This third stage brings alternating remission and relapse, with bouts of muscles aches, fatigue, depression and impaired memory and concentration. Old symptoms continue, and new ones arise. Arthritis, sometimes severe, strikes at least half of third-stage patients, attacking major joints such as a knee, hip or jaw. Inflammation of the brain and cranial nerves may worsen, causing mood disorders and the facial paralysis of Bell's palsy. In some European patients, the "tissue paper" skin lesions of ACA appear. In the worst cases, heart arrhythmia requires a pacemaker, and inflamed blood vessels invite strokes. Continuing insult to the nervous system can cause painful, numb or weakened limbs, double vision, and disturbances of thought or speech.

Many of these symptoms result from cascades of cellular and chemical events that start as soon as Bb enters the body. In a truly astonishing performance, the immune system promptly notices the invader, identifies it and summons up its frontline defenses. To do so it must distinguish Bb from the host's own cells and from countless microbes and poisons, including closely related spirochetes. This depends on mechanisms that have evolved over hundreds of millions of years, since invertebrates and plants developed the first rudimentary immune reactions. Our more complex system can sort through and identify 100 trillion different molecules and produce antibodies against them.

The body's network of defenses against Bb, too complex for more than the briefest suggestion here, starts with local inflammation around the tick bite. This increases the local blood supply, bringing in germ-fighting chemicals and specialized white blood cells that recognize intruders and signal for help. Actually, the white cells react not to Bb as a whole but to flagellin, Osps, PG, heat shock proteins and other surface molecules that act as antigens (organic substances provoking immune responses). As Bb travels on from its beachhead at the tick bite, the host's systemic defenses are kicking in. Other kinds of white cells go to work, each with special abilities—B cells, T cells, natural killer cells, engulfing phagocytes and others. The body also replies to Bb's surface molecules by creating an array of antibodies, proteins that attach to antigens and neutralize or destroy them. First there are short-term antibodies, an emergency response that is strong but not always selective enough to stop the invader. Later there are long-term antibodies, more finely matched to Bb's features.

It is these defenses, as much as Bb itself, that make people feel sick. Bb, we saw, does not poison hosts with toxins or cause messy symptoms to aid its transmission, such as coughing or diarrhea. But humans, like other mammals, react to most bacterial infections with fever, pain, fatigue and loss of interest in food, sex and grooming. That is primarily because several of Bb's outer-coat proteins trigger the release of potent hormone-like substances called interferons and interleukins. It is they that directly or indirectly cause fever, aches and malaise. For instance, interleukins turn up the brain's body-temperature thermostat and stimulate the production of prostaglandins, substances that provoke pain and inflammatiion. (Aspirin reduces fever and pain because it blocks prostaglandins.) Such symptoms may make life hard for the host, but they make it harder for the germ. For instance, fever kills some bacteria, inhibits growth in others, and steps up antibody production. And the same processes that cause fever and pain also shift iron from the blood to the liver, starving the bacteria that need it for growth and reproduction.

These and other defenses form a complex, elegant system that usually overcomes Bb, if not in days then in weeks or months; helped by antibiotics, it does so faster and more reliably.* But in chronic cases, Bb makes the immune system

---

*Antibiotics are wonderfully selective, attacking bacteria but not the cells of the human body. Some kill bacteria directly, but more do not. One type weakens a germ's outer wall until it breaks down; another

falter, spin out of control or both. This sometimes happens despite earlier treatment with antibiotics. There is intense debate about whether Lyme disease is chronic because Bb outwits the immune system, making it miss its target, or because it somehow leaves the system racing in high gear after the germ is gone. This is not only a theoretical argument; different explanations of chronic Lyme disease suggest different courses of treatment. The debate has been joined by the many informed laymen who belong to organizations for Lyme disease patients.

One view, held by many doctors and their third-stage patients, is that symptoms continue only if spirochetes are still hiding in the body. Bb, they point out, is a remarkable survivor; it can be cultured from EM and ACA skin lesions even after ten years of illness. While conceding that it is difficult or impossible to find Bb in some patients' bodies, they offer possible explanations. Bb may be elusive because of its very thin ranks; it never forms large colonies in hosts. And even in small numbers it can inflame joints or the heart, directly or by setting off immune reactions so strong that they punish host as well as germ. Perhaps Bb survives early antibiotic treatment by hiding in places drugs and antibodies have trouble reaching, such as brain cells and the white

---

keeps the germ from making proteins, so it cannot grow; yet another keeps it from making new DNA, so it cannot reproduce. Germs, weakened or unable to multiply, are finished off by the host's antibodies and white cells.

cells called macrophages (tricks performed by some other germs that cause prolonged illness). Or perhaps Bb shifts to a tiny L form without a normal outer wall; that would deprive some antibiotics of their target, helping Bb to persist. If some of these ideas are correct, and Bb does remain hidden in the body during third-stage Lyme disease, the best treatment is a long, strong course of intravenous antibiotics.

Another view, held largely by researchers, is that many chronic symptoms are caused by the immune system pummeling the body after Bb has vanished. They point out that this can happen in some other chronic bacterial infections. A postinfection syndrome may result from the immune system not switching off, for reasons still unclear. Or perhaps Bb ignites an autoimmune reaction, so that the body's defenses attack the self. In support of the latter idea, researchers note that some chronic Lyme disease symptoms resemble those of such autoinumme ills as lupus, rheumatoid arthritis and multiple sclerosis. They also resemble two other puzzling conditions that sometimes seem to follow Lyme disease, chronic fatigue syndrome and fibromyalgia. Perhaps Bb can trigger autoimmune or other syndromes that are mistaken for continuing infection. And finally, some still undiscovered germs may cause Lyme-like symptoms, which would explain why some people who apparently have Lyme disease test negative for antibodies against Bb.

If chronic Lyme disease is a hyperimmune or autoimmune disorder, heavy doses of antibiotics are futile; treatment should focus on relieving symptoms until the immune system regains equilibrium. The debate between those who call for antibiotics and those who recommend symptomatic treatment is stubborn and strident. At stake are the health and comfort of many patients with real or imputed third-stage Lyme disease. So far neither side in this debate has presented evidence that overwhelms the other.

Something else may help explain the variety of Lyme disease symptoms from person to person and region to region. More than 300 strains of Bb have been discovered worldwide, and a half dozen genetic subtypes. More are being found every year. These differences in genome and surface proteins probably account for ACA being common in Europe but very rare in North America. They may well explain why arthritis occurs more often in late-stage American patients, neurological ones more frequently in European patients. We know that some strains of Bb are more virulent than others, and that some are less susceptible than others to certain antibiotics. It seems logical that different strains and "genospecies" could cause different symptoms.

The existence of so many varieties of Bb has been a hurdle to developing tests, drugs and vaccines against it. It may also explain why some people catch Lyme disease a second time; immunity to one strain may not confer immunity to all others. Prevention and treatment are further complicated by the fact that several strains of Bb coexist in some regions, and sometimes inside a single tick. There are people who have been infected by as many as four strains of Bb at once. We will not fully understand Bb and Lyme disease until we have a better picture of the germ's various strains and their places in local ecosystems around the world.

TWENTY-FOUR

Like Darwin's Finches?

IN SOME PARTS OF THE WORLD, people might call my picture of Bb's life and relationships parochial. They would be right, of course; I have focused chiefly on the region where Bb was discovered, which also happens to be where I have spent most of my life. Bb, however, is at home around the world. It flourishes where a temperate climate encourages woodland biotas, generally avoiding tropics, deserts and long, fierce winters. Despite local variations, its life cycle follows the same basic pattern everywhere, from North America through Eurasia—transmitted by related species of hard ticks, given a reservoir by mice and other small rodents, and hosted as adults by deer and other large mammals.

In part of its global range, Bb is rarely noticed, because it rarely makes people ill. In some such places, it infects only ticks that do not bite people; in others, Bb and its reservoirs exist in small numbers, so there is little spillover to humans. And in some regions people hardly intrude into Bb's realm of forest and scrub. There is, however, a distinct high-risk zone for Lyme disease, and it reaches around the world. In the United States it includes the Northeast (Maryland to Maine), the upper Midwest (Wisconsin and Minnesota) and the West Coast (from northern California to Canada). It extends over most of Europe, crosses Russia and the Middle East, and continues through China, Korea and Japan. Lyme disease has also been found in Australia and South Africa, and there are some unconfirmed reports from tropical Africa and South America.

Both the low-risk and high-risk parts of Bb's habitat are growing; the circumstances and the germ's infective cycle vary from one ecosystem to another. We saw that in the American Northeast and Midwest, the basic cycle includes the deer tick, white-footed mouse and white-tailed deer. It is a bit different on the West Coast, where Bb has two major vectors. One is the tick *I. neotomae,* which feeds on the wood rat and other small mammals; it very rarely bites people, but it keeps Bb circulating in local wildlife. The other, the western black-legged tick *(I. pacificus),* carries Bb from dusky-footed wood rats to Columbian black-tailed deer, and sometimes to people. In low-risk regions such as the Southeast and the south-central states (for example, Missouri), others ticks are suspected of carrying Bb or related germs, occasionally transmitting Lyme disease or infections much like it.

Soon after Lyme disease was first detected in California, in 1978, the West Coast was found to be a high-risk zone.

However, Bb infects fewer people there than in the East, and the rate of increase is slower. Apparently the reason is a common lizard a few inches long, the western fence lizard. Western black-legged ticks often feed on this little reptile, but its blood contains a substance that kills spirochetes in the tick's gut, before they can be passed on. Since the lizard is not infected, it cannot pass Bb to new generations of ticks; as a result, the germ is less widespread in ticks and their hosts. In the Southeast as well, infected ticks feed on lizards that may act as buffers between Bb and humans. Buffer hosts also exist in parts of Eurasia, perhaps moderating the rate of Lyme disease there.

It is not just variations in vectors and hosts that modify Bb's infectious cycle. Climate and vegetation also affect the germ, and perhaps people who share its habitat. For instance, in the Northeast's oak forests, a snowy winter or balmy spring (we are not sure which) can produce a bumper crop of acorns. In a domino effect like that of the Machupo virus in San Joaquín, the acorns offer a feast to mice, whose population swells with their food supply; the mice nourish a thick crowd of ticks, which in turn host abundant borrelia. In years after big acorn crops, there are eight times more deer tick larvae than normal, and mice are heavily infested. There is no hard evidence yet of a direct link between acorn abundance and Lyme disease, but logic suggests that one exists. We can only guess how many other changes in climate, flora or fauna affect Bb's prevalence in a given place.

Europe, except for its extreme north and south, is also high-risk territory. There Bb is carried by the sheep tick, *I. ricinus,* from a reservoir in wood mice and other rodents. This tick ranges from Ireland to Iran and from southern Scandinavia to northern Spain; it also inhabits the forests of North Africa and the Near East. Bb and its vector have had more time in Europe than in North America to adapt to people and their land use; over millennia the sheep tick has become even more versatile than the deer tick. In some places it prefers woodland, in others scrub or pasture with coarse grasses; it feeds not only on mice and deer but on birds, lizards, cattle, horses and especially sheep. In Asia as well, Bb and its tick have long been adapting to human activity. There the germ inhabits the taiga tick, *I. persulcatus,* which carries Lyme disease from the Baltic states to the Pacific. Its range overlaps with the sheep tick's in Eastern Europe, and like its European relative, it has learned to feed on domesticated as well as wild animals.

In the Old World as in the New, Lyme disease is increasing in frequency and spreading geographically. Two decades ago only hundreds of cases were known to occur each year in the world's high-risk zones. Now more than ten thousand are reported annually in the United States, as many or more in Germany, and a thousand or more in Sweden, France, Switzerland, Austria and Czechoslovakia. Since the diagnosis and reporting of Lyme disease are uncertain everywhere, these numbers are unreliable. Some experts say that Lyme disease is overdiagnosed in the United States, while others think there are actually five or ten infections for each one reported.

Some reasons for the increase are clear. Everywhere some of the people at highest risk are loggers, hikers, hunters and children playing out-of-doors; in China, for instance, clusters of Lyme disease were first detected in forest workers. Many Europeans, like many Americans, are at risk because of reforestation and suburban sprawl. However, burgeoning deer herds and suburbs do not fully account for Bb's geographical spread. The expansion of Bb-infected ticks through the Hudson and Connecticut river valleys coincides not only with deer herds and human habitation but with the flyways of migrating birds. Tick-infested birds may have carried Lyme disease from the East to the upper Midwest, and perhaps along a coastal north-south axis as well. If this is so, birds may also be expanding the territories of Bb and its tick vectors in other parts of the world.

Before too long, we will probably be better able to describe Bb's life patterns around the world. At present we still do not know which tick carries Bb in Australia; nor which local variations in surface proteins make Lyme disease symptoms vary; nor whether the national and international transport of horses and dogs are expanding Bb's range; nor whether reports of Lyme disease in unlikely places result from yet undiscovered relatives of Bb. I would not be surprised to learn someday that many strains of

B. burgdorferi, with their varied host relationships, reflect local adaptations as neatly as Darwin's finches.*

It is not just intellectual curiosity that makes researchers want to understand Bb's adaptations; the knowledge may help them control Lyme disease. At first glance, the solution seems simple. Bb has many hosts, but it survives only if there are enough deer for infected ticks to mate on. If you eliminate deer, you eliminate the ticks, and with them Bb and Lyme disease. But sharply reducing deer herds, while physically simple, often proves to be socially impossible. Many farmers and homeowners loathe deer as pests and carriers of disease. Motorists fear them; thousands are injured and dozens killed each year by colliding with them. Yet many people who hate deer cannot bear the thought of killing them with gun or poison. You might as well suggest that they boil their children. And where some of them do accept the idea, their neighbors may not; communities often divide bitterly on the issue. When animal-rights advocates weigh in, action becomes impossible.

There are alternatives to culling herds, such as anti-tick pesticides, but these often arouse as much visceral opposi-

---

*Darwin studied the finches of the Galapagos Islands, those tortured rocks in the Pacific the Spanish called the Encantadas, or Enchanted Isles, and Melville described as "five-and-twenty heaps of cinders" scattered in the sea. On each isle the finches had a different food source, developed a unique bill to obtain it, and became a distinct species, unable to breed with the others. Darwin's finches are still the paradigm for how species emerge in ecological niches.

tion as deer kills. Many people fear that any pesticide will create a "silent spring" effect, devastating the environment. Such blanket resistance to pesticides has allowed malaria, dengue and other mosquito-borne diseases to threaten suburban Americans as well as the Third World's rural poor. Mindful of these anxieties, scientists have tinkered with biological controls, such as releasing minute wasps that feed on ticks and lacing deer feed with oral contraceptives. So far, their attempts have been ineffective or too expensive.

Public-health workers are stuck between low-tech solutions that people reject and ingenious ones that fail or cost too much. While awaiting better solutions, they warn people at risk to beware of ticks in woods, gardens, parks, school grounds, churchyards and the roughs of golf courses. They also exhort folks to inspect themselves daily for ticks, wear protective clothing and use tick repellents. Such vigilance does reduce Lyme disease but cannot eliminate it. Most people's watchfulness sometimes flags, and even conscientious inspection can miss the tiny nymphal ticks that usually pass Bb to humans.

A human Lyme disease vaccine became available in the late 1990s; it works by triggering the production of antibodies to Bb's surface protein OspA. This human antibody then works rather as the fence lizard's does; before a feeding tick can transmit Bb in its saliva, the antibody travels in the host's blood to the tick's gut and kills spirochetes there. This vaccine, like one approved earlier for dogs, is helpful, but it requires three injections over a year's time and is only about

80 percent effective. It will probably be used mostly in high-risk areas, and even there not everyone will be willing to get it or be able to pay for it.

Just as antibiotics do not always cure Lyme disease, these vaccines will not always prevent it. Eventually drugs and vaccines will doubtless be improved, but there still will be no flawless, universal barrier between us and *B. burgdorferi*. Diseases with nonhuman reservoirs cannot be eradicated, only held at bay. As long as the germ lives on inside one wild host somewhere in the world, it has a chance to reach humans again. Surely Bb, long ensconced in hundreds of reservoir and host species around the world, is safe for the indefinite future.

TWENTY-FIVE

A More Hopeful Future

THIS BOOK BEGAN BY SAYING that we can invent no drama more intricate and awesome than the smallest creature's biography. Even this brief sketch of a brief life has had to include a grand array of perspectives. In space they range from protein molecules to entire forests and continents; in time they vary from chemical cascades that flash by in nanoseconds to evolutionary trends slower than glaciers. Envisioning Bb's life and relationships can be like looking at once through a microscope and a telescope. Such complexity may make tidy minds fret, but to others it is part of the subject's appeal.

We have seen that change in a single species' life can reverberate throughout its biota in unforeseen ways. Bb is one example, humanity the ultimate case. Though our knowledge of Bb is new and incomplete, it already shows that we have repeatedly changed the germ's life by changing our own. Before we met Bb, it had spent thousands, perhaps millions of years evolving toward stable coexistence with its hosts. Then in some regions, in one evolutionary blink, we nearly wiped out its hosts and habitat. No sooner had we done so than we invited it back, helping it thrive and spread; we even became hosts ourselves. Today, having just begun to grasp this history, we are laboring to get Bb out of our lives again, or at least make ourselves immune to its infective power.

At this point one can only speculate what lies ahead for Bb and for us. I approach the task with misgivings, for people wiser than I have made such speculations and sounded very foolish soon afterward. Part of the reason is temporal provincialism, our difficulty imagining times unlike our own. Also, our vision of the future, as of the past, reflects not only evidence but fears and longings. Perhaps that is why so many prophets foretell utopia or dystopia, heaven or hell on earth, though so much of life really consists of anticlimaxes and muddle. Now, having invoked the future and excused myself in advance for getting it wrong, I must make my own guesses.

Humanity's future, I think, is more in doubt than Bb's. In fact, it is at hazard, and sometimes we seem to want it that way. The prevention of Lyme disease is a good instance. We would lessen human suffering if we reduced Bb's population and range, yet many people passionately oppose our best means for doing so. Steeped in ecological romanticism, they resist even judicious local use of anti-tick chemicals and limited culling of deer herds. In doing so, they engage in what the great researcher Peter Medawar called Arcadian think-ing, the notion that modern technology alone has despoiled and scarred nature. By nature they actually mean a world unmarked by human activity.

But in reality, no pure, unpeopled nature has existed for a very long time. About forty thousand years ago, paleolithic hunters improved their skills at hunting and at killing, and they went on to drive mammoths and other big game toward extinction. Since that leap to the top of the earth's food chain, we have not stopped changing our ecosystem more than it changes us. After the first super-hunters came neolithic tools, agriculture, metallurgy, urban life, industrial and technological revolutions, world travel and trade. It is not at all new that we change our biota and environment; what is new is how fast and radically we do so, sometimes with results that rightly frighten us.

Arcadian fantasy also ignores the fact that the human imagination and its consequences are themselves part of nature. They reflect the daring survival strategy of replacing much hard-wired behavior with learning and inventiveness, chiefly through vast expansion of the cerebral cortex. The score is not yet in on this unique and high-risk evolutionary gamble. It has made us equally ingenious at saving and destroying life. Unfortunately, it has not made us more adept as family, friends and neighbors, nor better able to foresee the effects of our inventions. It remains to be seen whether our gifts will someday do us in, perhaps taking other species or the whole planet with us.

Despite evidence that this is quite possible, I am cautiously optimistic. My attitude probably reflects in part a lifetime of listening to jeremiads—some right and some quite mistaken. In my childhood, fifty years ago, some scientific experts warned of an impending Malthusian disaster. The world, they said, could never feed many more than its then two billion people, yet its population was soaring. Besides, they added, in twenty years the earth would run out of oil. We were all destined to freeze and starve in the dark, if nuclear disaster did not kill us first. In the following decades we often heard that we were fated to choke on our wastes and see the last poisoned songbird fall from the last irradiated bough. Now we are told that our planet is becoming a foul greenhouse where Manhattan will sink into the sea and tropical parasites ravage Minneapolis.

If much of this doom-saying has failed to come true, it is not because the risks are imaginary. Pollution, radiation, global warming and new epidemics are very real threats; I need hardly recite a full litany of our environmental mischief. Yet so far we have survived some of our worst possibilities. New technology and occasional good sense have eased some of the problems we created: not all human tampering is for the worse. Many of today's six billion people are longer-lived, better fed and more prosperous than their grandparents were. Their very existence shows that we often

---

*Any opinion on such matters invites criticism. In another book, trying to be accurate and fair, I pointed out that the existence and extent of

fail to anticipate the good as well as the poisonous fruits of our behavior.*

So from a rational point of view, it seems moot whether humanity has a good future or even much future at all. But it is also moot whether we are racing toward self-imposed ruin. If I manage to see our future with sporadic hopefulness, it probably reflects not only my experience but my temperament. Like Dr. Johnson's friend who almost became a philosopher, I find that moments of good cheer keep breaking in upon me.

Borrelia burgdorferi's future seems more certain, and more promising. Right now the species is in very good shape in temperate regions the world over. And if our historical picture of the germ in the American Northeast is more or less correct, we have created there an enhanced version of Bb's ideal "edgy" environment. Bb probably lives there in greater numbers than ever before. There is nothing like a microbial census to confirm that, but recall that even in high-risk zones, not all mice and deer ticks are infected, and most deer ticks never bite people; at worst, only about five percent of deer tick bites transmit Lyme disease. This implies that for each reported case, there are legions upon legions of borrelia in an ecosystem.

---

long-term global warming are still debated by scientists, though the possibility demands both attention and action. One otherwise cordial reviewer took me to task for trying to use scare tactics. Another, also otherwise cordial, scolded me sternly for not scaring readers enough. So it goes.

Bb's population, the incidence of Lyme disease, and the size of high-risk regions are all increasing. They will probably keep rising as we create more woods, wildlife refuges, parks, suburbs and semirural dwellings. This is true not only in North America but in most of Europe and parts of Asia. Climatic change may also affect Bb's population and distribution. The germ has recently been found farther north in Sweden than ever before, in places once too cold for it. Some scientists explain this by local climate shifts, others by global warming. If the latter are right, the northern border of Bb's habitat may keep rising.

Bb and Lyme disease will also probably increase because of some things people will not do and some they cannot help. America is not scared enough of Lyme disease to substantially reduce the number of deer in high-risk areas. You can make a suburban home safer by fencing deer out; but even fencing has opponents, and there remains a riskier world beyond the fence. Most of us do not want to lose such secondary reservoirs for Bb as squirrels, chipmunks, raccoons and ground-feeding birds. We can greatly reduce the number of ticks and rodents in small areas such as lawns and urban parks, but not in large regions. In some high-risk areas, multiple efforts—tick-killing chemicals, tick repellents, protective clothing, body inspections, short-cropped lawn—give fair protection. However, they entail more sustained expense and vigilance than many people can manage. Still other steps might be needed in Europe, where Bb's main vector is the sheep tick, and in Asia against the taiga tick. And even despite all these controls, Bb will probably survive in small to moderate numbers, especially in low-risk regions. Having lived for so long outside the sphere of human life, it will doubtless continue doing so.

Right now biomedical science holds little danger for Bb. Soon there will probably be better vaccines and drugs against Lyme disease, but they will affect people more than the germ. Even if Bb should lose humans and many domesticated animals as hosts, its life will change little; none of these species are reservoirs that sustain its infective cycle. We and our pets and herds have never been more than a sort of biological overflow basin when Bb is plentiful and we invade its realm.

The things that might seriously affect Bb are the same ones that did so in the past—a major die-off of deer, unforeseen changes in land use, and altered human behavior. At present these are more likely to happen by accident than on purpose. Even if they did, this compact marvel of adaptation would probably show again how well it can survive as a species. Short of a doomsday nuclear event, it can probably outlive anything we can throw at it. In fact, as I try to imagine what might truly bring an end to *Borrelia burgdorferi*'s long history, I can picture only one of those cataclysms that at long intervals threaten much life on earth—the crash of asteroids, great heats and freezes, a changed atmosphere, the rise and fall of seas. Heaven knows what these events would do to people. But in my mind's eye I can see emerging from the remaining urban rubble or flattened forest a small wild mouse. On the mouse is a tiny tick, and in the tick there lazily turns a minute, slender spiral, hardy and elegant, again a survivor when so many creatures have had their day and gone.

About the Author

Arno Karlen, Ph.D., is a psychoanalyst who has written on the history of medicine and biology. His essays on literature, history, medicine and behavioral science have appeared in many scholarly and popular magazines. He lives in New York City.